"A REMARKABLE STORY OF ONE MAN'S PRIDE, COURAGE AND DETERMINATION . . . WE ARE ALL THE BETTER FOR IT."

—Armand Hammer, CEO, Occidental Petroleum

"An important biography because one sad consequence of the media's total identification of African-American history with blacks who are poor and powerless is our frequent inability to see or fully appreciate proud, self-sufficient nationalists like Jake Simmons, Jr. . . . the author provides wonderfully detailed research on the 19th-century relationship between blacks and Indians. . . . Few men seem destined to shape history quite as thoroughly as did Jake Simmons, Jr."

—*Los Angeles Times*

"A marvelous story of a fascinating family, superbly told."

—David J. Garrow, Pulitzer Prize-winning author of *Bearing the Cross*

"A fascinating story . . . a must addition to . . . African-American, Indian, and U.S. history."

—*Booklist*

"Stakes a new claim for the continuing vitality of the American dream . . . transcends race and color and should be read by all Americans."

—John H. Johnson, publisher of *Ebony* and *Jet*

JONATHAN GREENBERG is a free-lance journalist and former *Forbes* reporter whose work has also appeared in *Manhattan, Inc.*, *New York* magazine, *The New York Times*, and *Mother Jones*.

STAKING A CLAIM

Jake Simmons and the Making of an African-American Oil Dynasty

JONATHAN D. GREENBERG

A PLUME BOOK

PLUME
Published by the Penguin Group
Penguin Books USA Inc., 375 Hudson Street, New York, New York 10014, U.S.A.
Penguin Books Ltd, 27 Wrights Lane, London W8 5TZ, England
Penguin Books Australia Ltd, Ringwood, Victoria, Australia
Penguin Books Canada Ltd, 2801 John Street, Markham, Ontario, Canada L3R 1B4
Penguin Books (N.Z.) Ltd, 182-190 Wairau Road, Auckland 10, New Zealand

Penguin Books Ltd, Registered Offices: Harmondsworth, Middlesex, England

Published by Plume, an imprint of New American Library, a division of Penguin Books USA Inc. This is an authorized reprint of a hardcover edition published by Atheneum Publishers.

First Plume Printing, January, 1991
10 9 8 7 6 5 4 3 2 1

LIBRARY OF CONGRESS CATALOGING-IN-PUBLICATION DATA

Greenberg, Jonathan D.
Staking a claim : Jake Simmons and the making of an African-American oil dynasty / Jonathan D. Greenberg.
p. cm.
Reprint. Originally published: New York : Atheneum, 1990.
1. Simmons, Jake, 1901-1981. 2. Afro-Americans in business--Biography. 3. Petroleum industry and trade--United States--History. 4. Petroleum industry and trade--Africa, West--History. 5. Civil rights movements--United States--History. I. Title.
[HD9570.S56G74 1991]
338.7'6223382'092--dc20
[B]
90-46195
CIP

Printed in the United States of America

Contents

STAKING A CLAIM

Introduction

I first heard about Jake Simmons in early March, 1982. I was stopping in Tulsa during a year-long research tour from which I was to emerge with a list of the four hundred richest people in the United States for the very first *Forbes* "Rich List." Malcolm Forbes had told the editors to produce a roster of the ultra-affluent so people could talk about the "*Forbes* 400" the way they talked about the "*Fortune* 500." Finding four hundred Americans with a net worth of more than $100 million was not easy, especially since my initial instructions were to make sure that those people we were eventually going to name didn't know it until it was too late for them to do anything (like sue). This meant that for a long time I couldn't approach the super-rich themselves; I had to slink around and talk secretly to the very rich who knew them.

That's what brought me to Taft Welch, the senior board chairman of the Western National Bank in Oklahoma. Welch is a typical frontier success story. He went to Tulsa in 1923, a tall, lanky fifteen-year-old eager to make his fortune. When he arrived the Ku Klux Klan was at the height of its popularity. He told me he had won a Klan concession to sell members of the hooded order inexpensive white robes for sixty dollars each "at a time when dress suits cost fifteen dollars." Quickly he graduated into trading bankrupt companies, then invested in oil properties. Although he recalls that he was just "a raggedy-assed kid" when he made his first million, he warned his eleven sons "if they marry money I'll cut their throats—the poorer the better."

When I met Welch he was seventy-five. He received hourly phone calls from a broker and dictated hundred-thousand-dollar trades while being interviewed. We talked across a mahogany desk big enough to land a helicopter on. There wasn't a scrap of paper on it; clearly it was the desk of a boss too old to be bothered with the trifling details of paperwork. Welch was hospitable and talkative. His nose was bright red, as was his chafing, almost bald head, and he spoke with a heavy southwestern drawl. The only object I noticed in his desk was a round tin pail in an open bottom drawer. He chewed tobacco as we spoke and every two minutes or so he would abruptly clear his throat and, only half-bothering to bend over, spit a huge ball of murky phlegm into it.

Did he know any local centimillionaires? I asked, showing him a stack of "suspect" cards I'd worked up. Of course he knew them—he knew them all, and what he didn't know he was more than happy to guess at. The rich list was inherently a guessing game anyway, so Taft Welch was the perfect source.

"You're missin' Jake Simmons from Muskogee," he said, after commenting on those mentioned on the cards. "The colored fella who put Phillips into Africa. He's a born trader—worth more than any of the names you got heah!"

I never managed to figure out how much money Simmons was worth, although I am now certain Welch was making a pretty wild guess, and that while he was successful he was never Rich List rich. The nature of Simmons's business made it impossible to assess his income, and the secrecy of his family's multinational operation was unbreachable. I did, however, find out many other things about Simmons and his family. And this, I discovered, said a great deal more than an unattainable number.

A week after talking to Taft Welch, I telephoned Donald Simmons, the son of Jake Simmons, Jr., who had died recently. I asked if I might write a *Forbes* profile about his family's oil-trading enterprise in West Africa. At first I received a flat refusal. Donald's father had built his fortune and influence as a power behind the scenes. He impressed upon Donald (the son who took over the family oil business) the need to keep a low public profile—especially regarding their affairs in Africa. Talking about the past can be a sensitive subject in a region so politically volatile that last year's cabinet members often turn out to be this year's condemned convicts. (I came to appreciate this aspect of African life a few years ago when, after I left Ghana, a young man I had asked to expedite a request for innocuous information from a government agency was interrogated as a spy.) No member of the Simmons family had ever been written about in a national publication and only once, in 1963 (when the *Daily Oklahoman* profiled Jake Simmons, Jr.'s, role as president of the state's NAACP), did a daily newspaper make more than a passing reference to the Muskogee oilman. When I suggested that *Forbes* might like to portray them as one of the most successful black families in America, Donald said, "No way—even if we had it, we wouldn't own up to it anyway. My dad would roll over in his grave if you said that."

I discussed with Simmons the notable absence of black businessmen in financial magazines like *Forbes*. I stressed the importance of commemorating the achievements of his recently deceased father, an unsung entrepreneur who not only became the world's first internationally successful black oilman, but also managed to integrate into his life a bold career as a civil rights leader. Months later, after consulting with other members of his family, he agreed to an interview in Muskogee.

I met Donald Simmons for breakfast at the Holiday Inn on Muskogee's main strip at 8:00 A.M. on June 29, 1982. I wasn't sure what to expect. I didn't think he would bear much resemblance to those I had become accustomed to interviewing for *Forbes,* the lean, well-polished corporate executives who wore conservatively tailored suits and oozed public relations charm. I wasn't disappointed. Simmons stood six feet one and wore a cowboy hat, western boots, and a gold Rolex watch. He carried his 250-pound frame like a man to whom eating voraciously was just one aspect of living a robust life. It took him less than ten minutes to liberate his body from a casual sports jacket. He knew the waitresses by name and knew what I should order and knew what was wrong with modern America. He turned out to be one of the most articulate men I had ever met.

After breakfast we drove in his Jeep out to a small oil rig, where I interviewed two white drillers who had worked with the Simmons family for decades. We then moved to the quiet, well-appointed office of the Simmons Royalty Company, in the heart of downtown Muskogee (a sedate city of 40,000 so conservative that a copy of *Playboy*—even behind the counter—is harder to find than a slice of quiche). Our conversation continued through dinner at Slick's barbecue, where Oklahoma's best ribs are slapped onto wax paper and served with two pieces of Wonder Bread and a raw onion, and no one but the boss is allowed to handle the money or cut the ribs. Simmons introduced me to Slick, the seventy-year-old gray-haired black proprietor and local legend. Slick showed me the pearl-handled .45 Colt revolver he kept close by, and gave me a tour of the hickory pit out back where he smokes hundreds of pounds of ribs every day (except when he closes up shop to go fishing).

Finally, in a spacious home decorated with African carvings, where Simmons lived with his family, we concluded our first interview at 11:00 P.M. For fifteen hours Donald Simmons had proceeded like a talking dynamo, and there was scarcely a moment when I wasn't captivated by the conversation.

We spoke mostly of Donald's father, Jake Simmons, Jr., who died in 1981. The most successful African-American in the history of the oil industry, Jake Simmons made a name for himself in one of the most conservative businesses in the United States. He was a man who believed that life was meant to be a continuous struggle, a challenge to be met by working a little harder than was necessary, by carrying himself one step further than what was required, and by proving beyond all doubt that the actualized power of his potential was greater than any measure of racism which sought to restrain him.

Jake Simmons was born in 1901 to the granddaughter of Cow Tom, one of the only black chiefs ever to lead a Native American tribe. Starting in Indian Territory, an exceptional bastion of black dignity, his life spanned the entire civil rights struggle of the twentieth century. Born one of fourteen children on a thousand-acre ranch, he grew up as part of the Creek tribe in a black Indian community which had prospered beyond the range of racist America's state or federal government. Since the end of the Civil War blacks had been allowed full economic and political participation in their tribe, and had achieved a level of enfranchisement unknown to African-American citizens of the United States.

When Jake turned seven Oklahoma became a state, displacing the tribal government. Just as state governments across the South rolled back Reconstruction in 1877, Oklahoma's new white rulers immediately tried to legislate second-class citizenship for their black constituents.

It didn't work. Tribal citizenry had built forty years of freedom and self-esteem into black Creek culture. Bigoted attempts to tighten the screws of racial barriers through whippings and lynchings resulted in the Tulsa race riot of 1921, the bloodiest in the nation's history. Because people like Jake Simmons were willing to lay down their lives to uphold their dignity, they were able to win much of the respect they fought for.

Jake Simmons fought for a lot. His father would have liked him to run the family's cattle ranch, but he was enamored of the exploding oil industry. He knew what he wanted from an early age, and it was not farm life; as a teen he was so embarrassed to be seen in overalls that when girls his age came to visit he hid out in the fields until they left. A visit to the Simmons home by Booker T. Washington in 1914 convinced his father that Tuskegee Institute was the place to educate his boys. Young Jake had the privilege of studying with the famous educator at the enclave of black self-sufficiency that was Tuskegee. He learned from a master the art of public speaking, and shaped his ambition at an institution which stressed hard work and pragmatism above all.

Simmons got married right after graduation to a northern woman, then tried working as a machinist at a Detroit auto factory. When his bigoted superiors refused to look at a defroster he invented, he quit the job, divorced his first wife, and headed back to Oklahoma, where he could settle with his hometown sweetheart and be his own boss. A driven fighter, he worked unceasingly to defy limitations and triumph over obstacles. Time and time again he refused to be a victim of bigotry, to suffer from preconceptions others imposed on him. "You are equal to anyone," he would tell his children, "but if you think you're not you're not." With this approach he took on the prohibitive stereotypes of a segregated industry, and won the respect of oil tycoons who had never regarded members of his race as anything but "boys." He started out in Oklahoma and Texas by brokering oil leases for black farmers wary of being ripped off by oil companies. He earned both their trust and a percentage of their oil royalties, then made sure the deals he brokered were honored.

Simmons's expertise as an articulate and persuasive salesman eventually allowed him to bridge the gap between postcolonial West African nations seeking outside investment and international American mineral companies looking for opportunities in foreign lands. Although he never studied geology in school he knew enough about the subject to make the chiefs of billion-dollar oil companies believe he held a master's degree in the subject. He explained international contracts to inexperienced high-ranking officials in Liberia, Nigeria, and Ghana, negotiating amiably where the world's largest oil companies, the so-called Seven Sisters, feared to tread. On most deals he retained an "override" percentage of all the oil that came out of the ground. He then peddled shares of his piece of the action, assuring himself of a profit regardless of how much oil was found.

The foreign officials Simmons dealt with came and went, but he had an uncanny ability to land on the good side of whoever was in power. Newborn African states like Nigeria and Ghana had only shaken free from the British Empire for a few years when Simmons arrived there in the mid-1960s. The fledgling nations were as eager to shed their dependency on Britain's petroleum giants (British Petroleum and Shell) as they were to be free of her political domination. The natural choice was to seek investment from America's multinationals, but the inexperienced officials were not sure whom to trust.

During the sensitive Nigerian negotiations for what proved to be the most profitable oil deal an American company ever struck in sub-Sahara Africa, Simmons helped bridge this distrust. A black nationalist with strong political opinions, Simmons was welcomed as a person whose loyalty and honor did not end outside his corporation's boardroom.

Somehow Jake Simmons managed to remain his own man. He was a frenetic intermediary, setting up deals between huge multinationals like Phillips Petroleum, Texaco, and Signal, and top government officials from Nigeria, Ghana, Liberia, and the Ivory Coast. He worked for big oil and he worked for big governments, but he let everyone know that in the end he worked for Jake Simmons. He believed in capitalism and the opportunities of the American system, and was determined to make those opportunities exist for himself and those of his race. He knew how to spot possibilities when others saw only despair. Empowered by what he called "bulldog tenacity," he was willing to sweat through a dozen bum deals before one panned out. Simmons personified the American entrepreneurial spirit. "One-tenth of the folks run the world," he would tell close friends. "One-tenth watch them run it, and the other eighty percent don't know what the hell's going on."[1]

Africans respected him for his fierce sense of racial pride. He was a great talker. In a land where ancestry is revered, he told incredible stories of his black Indian forefathers. Beyond being a man of his word Simmons left African associates with a sense that when push came to shove he was on their side.

Only once did he have to prove this. It was at the end of 1968. Simmons had spent nearly two years assisting the Ghanaian Ministry of Lands and Mineral Resources in the development of its first comprehensive petroleum investment code. At the same time he had interested an American consortium of Signal Oil, Occidental Petroleum, and Amoco in backing an exploration deal he was assembling in Ghana. Some of the world's largest oil companies, including Exxon, Gulf, Texaco, and Mobil, had also negotiated exploration deals with the Ghanaian government. All that was left was the contract-signing ceremony. The night before the ceremony an American diplomat appeared at the home of oil commissioner Sylvan Amegashie and warned that the American multinationals had met privately and would refuse to sign. The cartel was looking to renegotiate the conditions of the petroleum code.

The moment the diplomat left Amegashie called Simmons, the man he remembers as his "mentor." Amegashie was nervous. "What's this I hear about the Americans forming a conspiracy to break this thing up?" he asked.

"Rest assured," Simmons replied, "our consortium will sign." The next day, at a lavish ceremony in the Ghanaian State House, Amegashie made a speech, then invited representatives to the dais to sign their concession contracts. Simmons jumped up to sign, but not before making a speech of his own. "Ghana is a country struggling to emerge from underdevelopment," he said, "and the rest of the developed world should make it its business to respond to this."[2]

The cartel's hand was broken. Over the next decade American multinationals invested more than $150 million in Ghana. Much of it came from deals Simmons brought in. Just three years before his death, in a 1978 ceremony attended by hundreds of thousands, Ghana's head of state awarded him the Grand Medal, the nation's highest honor.[3]

Back home in Muskogee, Oklahoma, Simmons's honors were less grandiose but no less significant. Because he was his own boss the oilman was uniquely independent from the system of political patronage, and became a force to be reckoned with in eastern Oklahoma. Powerful politicians wooed and tried to bribe him; his opinion was said to have been able to sway 10,000 voters. In 1937 he and his wife initiated one of the earliest school desegregation cases ever to reach the Supreme Court. Death threats and an organized effort by Muskogee's power brokers to run him out of town were met by a solemn determination and an understanding that his home was well stocked with firearms. A strong faith in God led to a belief in his own destiny, and a fearlessness of mortal men. "You need God in your life to get anywhere," he would say, then quote Scripture. For more than forty years he was president of the laymen's association of Muskogee's large African Methodist Episcopal church, and influential in the church's national affairs.

Simmons didn't need a church pulpit to deliver a sermon. At barbecues and school meetings, and on street corners, he never forgot Booker T. Washington's secrets of public speaking, and drove terror into the hearts of racist politicians whenever he took the "stump." As leader of Oklahoma's NAACP, he led the fight against segregation and discrimination, supporting the nation's first lunch counter sit-in effort while the area's more conservative black leaders argued for moderation. "Full citizenship rights NOW," he wrote, "moderation LATER."[4]

To Simmons, jobs were the key to the economic empowerment which far too many African-Americans lacked. As he climbed the ladder of success he never stopped looking back. "For a man not to achieve the things he has an opportunity to do is such a waste of life," he said. "Not for selfishness, but for what you can do for others." Simmons applied pressure on politicians to hire blacks, and saw to it that hundreds found jobs. African-Americans would line up for hours outside his office, where he offered financial and personal advice like a local godfather. His favorite charity was people. "Anybody in the world could get fifty dollars out of him with a hard-luck story," one of his sons noted. "They just had to sit down and let him lecture them."[5]

Despite his political power, and the fact that he was among the first black delegates ever to represent Oklahoma at a Presidential convention, Simmons never won a single election. Three times he ran for local office, and three times he lost. Blacks turned out in record number to vote for him, but white votes—essential for the at-large seats he contested—were scarce. Politically he was the very sort of African-American the white power structure did not want on the inside. Stern, obstinate, and straightforward, he never liked to compromise. He told people what he thought in no uncertain terms. Although he supported the concept of nonviolence, he found it personally untenable, and once had to be physically restrained during a nonviolent protest in which a white bigot antagonized him. (His well-known willingness to resort to fists—or weapons—to defend himself kept many an assailant at arm's length.) Unelectable, he made his power felt behind the scenes, commanding more respect out of office than any of the state's blacks who eventually gained office. Senators, governors, congressmen, and even Presidents vied for his support and advice. He used his influence to help promote his sons, who in turn helped advance his interests. Lyndon Johnson's former chief of staff called him a "black Joe Kennedy."[6]

Simmons tried to keep his business separate from his political activism, but this was not always possible. His fierce opposition to Oklahoma's powerful Senator Robert Kerr, "the uncrowned king of the Senate," cost him what might have been an enormous bonanza. During Oklahoma's 1954 senatorial campaign Simmons came out as a leading voice against Kerr because of his dismal voting record on civil rights. Then, in the mid-fifties, the Liberian government awarded Simmons one of the largest mineral concessions ever given to an individual, stretching over seven million acres—nearly half the West

African nation. He interested the heads of one large multinational firm in backing an exploration program, but they conditioned their involvement on the partnership of Oklahoma's Kerr-McGee Corporation, which Senator Kerr retained control of while in the Senate. Kerr squashed the deal and Simmons had to give back the concession.[7]

It was an expensive loss. But Simmons was used to taking his lumps. Despite his enduring optimism and relentless ability to persevere, he was never a particularly lucky man. The projects in which he held the largest pieces rarely worked out the way he hoped they would—but he always hedged his bets and looked starry-eyed to the next deal. The trick was never to let up. When he was sixty-four he wrote a letter to his sons which said, "I subscribe myself to the philosophy that mankind was made for struggle, for hardship, for enduring suffering. Such was the plight of our race, and who will challenge our eighty years of progress in America? Mankind grows strong and noble under strain."

The story of Donald and Jake Simmons's success in the African oil business, entitled "Mr. Two Per Cent," ran two pages in *Forbes* in the summer of 1982. It was one of the first profiles of a self-employed black entrepreneur ever published by the sixty-three-year-old magazine.

The short article I wrote proved to be the tip of an intriguing iceberg. The rest of the story and what I considered its historical significance captured my imagination. Much has been written on the various ways in which blacks have been kept down in America. Here was a series of exceptional circumstances which allowed one black family to rise, a true tale of a frontier territory which provided its African-American citizens an opportunity to partake in the American dream in a manner unknown anywhere else.

For more than a year after the *Forbes* article appeared, I discussed the idea of a book with Donald Simmons. He was noncommittal. He'd have to see what I had in mind. In mid-1983 I left my job and came back to Oklahoma to show him.

He was not an easy man to convince. My affinity for the subject was not immediately apparent from my skin color or geographic background. The trust between many members of the Simmons family and myself came not from any common experience but from the zeal with which I pursued the project. In the terms of the Oklahoma frontier, I became worthy of their confidence by homesteading my way into it.

The pursuit of the story took me into a world I knew nothing about. Rural Oklahoma, where I lived for more than a year, was a far cry from my Greenwich Village home. Neighbors are important in Oklahoma, and I found that hospitality—genuine hospitality, not plastic smiles—is alive and well in the American heartland. When I visited somebody I most often visited their family as well. I found myself delighted to count among my new friends octogenarians as well as infants. Yet to a progressive like myself there was also a disorienting dimension to living there: the paradox of finding that some of the same individuals who embraced ultraconservative politics also possessed enormous human warmth was enough to make my head swirl. The same retired Oklahoman who believed that a nefarious international Zionist conspiracy rules the planet (partly through the insidious "eastern media establishment" from which I hailed) was as willing to accept me as a trusted neighbor as he would his own grandson. Perhaps it is this very paradox, this inexplicable willingness of those on the frontier to accept in the flesh what they bitterly reject in principle, that partly explains how Jake Simmons was able to triumph in a world constructed to defeat him.

The history of the Simmons family between 1830 and 1980 is the history of a class of African-Americans whose resistance to racism was rooted in their unique Indian heritage. From such a foundation each generation reached the height of black power during their day. As I contrasted the experience of black members of the Creek tribe with that of blacks in the rest of America, I was struck by how very exceptional the social and economic opportunities which existed in prestatehood Oklahoma were. In the independent Creek nation, African-American tribesmen were protected from white America's racist backlash against Civil War Reconstruction. Black Indians from Indian Territory were the only citizens of their race to be permanently enfranchised into American society the way they were intended to be by those who made them equal in the U.S. Constitution. Many historians wonder what the economic status of millions of African-Americans would be today if blacks had truly been given their freedom after the Civil War. I hope the little-known story of Jake Simmons and his ancestors' success in transcending the limitations of their turbulent times can provide new insight into this critical question.

PROLOGUE: AN ANCESTRY OF FREEDOM

As there are among the Creeks many persons of African descent, who have no interest in the soil, it is stipulated that hereafter these persons . . . and their descendants . . . shall have and enjoy all the rights and privileges of native citizens, including an equal interest in the soil and national funds.[1]
—Creek Treaty of 1866, developed under the auspices of Cow Tom, great-grandfather of Jake Simmons, Jr.

Of all statehood-day celebrations, the only one to stand out in American history is Oklahoma's. Rodgers and Hammerstein immortalized the event in their famous musical, and Oklahomans are unusually proud of their frontier heritage. Many can even tell you what their relatives were doing that very day.

If the ancestor was white, November 16, 1907, probably found him or her in the street celebrating. On statehood day eastern Oklahomans poured into Muskogee, the town local businessmen dubbed the "Queen City of the Southwest." Banners streamed from twelve-story skyscrapers, firecrackers exploded, sparklers blazed, and train whistles wailed. Strangers hugged and men whooped. They had good reason to rejoice. For decades whites choosing to live in the so-called Indian Territory had submitted to the Indian law. With statehood white majority rule would join the backward territory to modern America. With statehood everything would change.

For the earliest settlers of Indian Territory, all of them nonwhite, it would be a change for the worse. African-Americans who were members of the area's Indian tribes, like Jake Simmons's father, knew that the statehood bells signified the end of an era. Although Muskogee was only fifteen miles away from his family's sprawling ranch, six-year-old Jake and his brothers and sisters were kept home for the holiday. They longed to be closer to the spectacle, but their father would not hear of it. The new state, he explained, was to be run by "rebels," racists committed to "putting blacks down." For two generations a few thousand African-Americans had found refuge from white racism by living among an Indian nation. That refuge had just been shattered. In its place stood the new state of Oklahoma.[2]

The story of Jake Simmons begins with Cow Tom, a man who carved a unique niche for black Indians on the American frontier.

Cow Tom started life as a slave in 1810. Unlike most slaves in North America he was owned by an American Indian, an affluent leader of the Creek tribe named Yargee. Since the late eighteenth century slaves had become a status symbol for wealthy Creeks, a sign that they, too, could enjoy the lifestyle of the white man. Tribesmen lived in houses, maintained a well-organized system of government (headed by a "principal chief" and nearly forty regional chiefs), and owned slaves. Some were purchased (the going price was $200), some were stolen, and some came to the tribe as runaways from the plantations of harsh white masters.[3]

Living on lush, fertile lands in what is now Alabama, the Creeks were a content people. Neither the heartlessness of unbridled capitalism nor the religiously supported doctrine of racial supremacy had much foundation in their culture. In *The Road to Disappearance,* a definitive history of the Creek tribe, historian Angie Debo noted, "Except on the plantations of a few mixed (white and red) bloods, slavery rested very lightly upon the Creek Negroes. The easygoing Indians found the possession of slaves a great convenience, but they saw little reason to adopt the white man's ruthless system of exploiting and degrading them."[4]

Yargee, Cow Tom's owner, never sold a slave. He provided each with decent housing and an individual plot of plush farmland near the banks of the Alabama River. The slaves were allowed to own property and even work on the side for money to purchase their freedom.

Intermarriage between the two races was common, and the offspring of such unions were well treated.[5]

Cow Tom had a choice job among Creek slaves. As a young man he tended cattle (hence the name, to differentiate him from other Toms). Among the Creeks the accumulation of money counted for little, and land was commonly held. In the early nineteenth century it was the size of a man's herd—and his war honors—which determined his prestige. Yet the Creeks were new to livestock-keeping. They had traditionally relied more on hunting and trapping than stock-raising. Because the encroachment of white civilization had decreased the supply of game, cows were seen as the wave of the future. The job of tending the fast-growing herds fell to a tribesman's best slave, who was thought to better understand the new ways of the white man's world than his master.[6]

Historically Creek men had built their reputations on their war skills. Proficient diplomats and superb fighters, the Creeks once held the balance of colonial power between the Spanish in Florida, the British in Charleston, and the French in Louisiana. Their mighty "Creek Confederacy" encompassed conquered tribes and numbered more than twenty thousand citizens, controlling thirty million acres (much of the territory that now makes up Alabama and Georgia). On the eve of the Revolutionary War a British Indian expert said the Creeks were "unquestionably the most powerful Indian nation known."[7]

In the early nineteenth century America's expansionist power changed all that. Federal officials began to regret the treaties they'd signed recognizing the borders of the region's Indian nations, and used every opportunity to find ways around them. The Red Stick Rebellion, a bloody uprising by a nationalistic faction of the Creek tribe in 1813, gave General Andrew Jackson an excuse to bring a huge army into the area and slaughter the tribe's fiercest warriors. (Afterward he prophetically gloated that the Creek Confederacy's power was broken forever.) Although a majority of the Creeks fought with Jackson against the rebellious Red Sticks, the general forced the tribe to give the government two-thirds of its land—twenty-two million acres—as a war indemnity.[8]

The land grab did little to appease the white man's hunger for Indian real estate. Hundreds of settlers from Alabama, Georgia, and

Tennessee obtained phony deeds, occupied tribal land, and plundered Indian property at will. State courts refused to accept the red man's testimony, and whenever tribesmen attempted to enforce order themselves they were overpowered by mobs of heavily armed white thugs who called themselves volunteers from the "state militia."[9]

Locals seized upon every incident to demand Indian deportation. They argued that the Indians should be shipped west, relocated to an unwanted section of land left over from the Louisiana Purchase. The Creek Council refused to swap their ancient homeland, and imposed a death sentence on any chief who agreed to sell tribal property.

In 1825 federal Indian agents bypassed official tribal channels and bribed a wealthy half-white chief $25,000 to authorize the exchange of the entire Creek nation in the Southeast for a large tract of desolate land where Oklahoma is now situated. The chief was promptly executed by a squad of one hundred warriors, but federal officials insisted that the swap was valid. In 1829 Andrew Jackson was elected President with the stated intention of driving the Indians west of the Mississippi. Congress bitterly resisted. Jackson's "removal" policy became the most hotly contested issue of his administration. The war hero prevailed. In 1830 the Indian Removal Bill was enacted into law.[10]

The uprooting of the Creek nation meant a sudden change in professions for Cow Tom. A well-built, jet-black man of average size, he was highly regarded by his owner for his sharp, opportunistic mind. Yargee used him as his personal interpreter on the rare occasions he needed to deal with whites. As the deadline of his tribe's forced departure from the Southeast approached he found a way to cash in on his slaves' abilities.

Before allowing the Creeks to leave their homeland for Indian Territory, President Jackson forced a regiment of 777 Creek "volunteers" to fight on the U.S. Army's side in the Seminole Indian War. The army needed interpreters to accompany them; Yargee was probably able to earn $350 by leasing Cow Tom to the war effort.[11]

The Seminole War, which raged in the swamps of northern Florida between 1835 and 1842, was the bloodiest Indian War in American history. The tiny Seminole tribe, an offshoot of the Creek nation, refused to cooperate with America's Indian Removal program and waged a guerrilla war which cost the U.S. Army $40 million and the lives of fifteen hundred soldiers.

Much of the reason for the Seminoles' refusal to move had to do with the hundreds of runaway black warriors living among the tribe. The Seminoles were the most benign black slaveholders in American history, and intermarried so extensively that a federal agent probably chose his words carefully when he noted, "An Indian would as soon sell his child as his slave." Black Seminoles encouraged the Native Americans' resentment for the encroaching whites. Because removal would have meant reenslavement for many black tribesmen, they were willing to risk their lives to fight emigration, and convinced their red brethren to do the same.

The U.S. government, meanwhile, was terrified that the black Seminoles would encourage slave revolts in neighboring white areas and create an army of black liberators. It poured a steady supply of federal troops and arms into the uncharted malaria-infested swamps of northern Florida, but they were helpless against the Seminoles' guerrilla tactics. American generals, convinced that their best allies in wars against Indians were other Indians, mustered as many Creek fighters as they could force to join.[12]

The Creek Regiment's first commanding officer, a white lieutenant-colonel to whom Cow Tom was assigned as interpreter, was driven crazy by the heat and committed suicide within a few weeks of arriving in Florida.[13] Cow Tom was reassigned to Major General Thomas Jesup, who was in charge of the entire Florida campaign. Of the forty interpreters used by the army for both its Creek allies and Seminole enemies, it was Cow Tom whom the general trusted most.[14] Cow Tom's stature grew with the task. In his book, *History of the Second Seminole War*, John H. Mahon described the importance of the interpreter: "Few white men could handle Hitichi and Muskogee; few Indians, English. Thus, the interpreters were the channels through which information had to flow. In most cases they were Indian-Negroes, devoid of education, yet what they reported was perforce the basis of all official action. The interpreters . . . even though they were usually slaves, were as important in any negotiation as the most exalted person present."[15]

They were also in a position—quite rare in wartime—to know

what was really going on. Cow Tom quickly discovered that the fighting was more over slavery than anything else. The only way that Jesup was able to get his enemy to sign a peace treaty (interpreted by Cow Tom) was by guaranteeing that all the tribe's so-called slaves would be able to leave Florida. When Jesup reneged on his promise and allowed slave catchers into the army-run Seminole emigration camp, the Indians began to plan a breakout. Jesup sent several Creek spies into the compound to inform him of any unusual activity. Cow Tom was probably among them. Presented with a conflict between race and duty he chose to do nothing. While two hundred Seminole warriors emptied the camp of its detainees, he and the other spies kept quiet.

Jesup blamed the Creek spies for the escape and began to question the loyalty of his reluctant Native American allies. By this time—mid-1837—all but thirty of the more than seven hundred Creek "volunteers" claimed to be too sick to fight. Jesup released the Indian regiment from duty.[16] They joined their families in refugee camps in Mississippi to begin their journey west, a forced exodus historians later dubbed the "Trail of Tears."

Cow Tom, his wife, Amy, and two young daughters were lucky. Unlike hundreds who succumbed to accidents, disease, and starvation on the wintry trail (including one of Yargee's sons), they all arrived alive at their new home in the West.[17] And although Indian Removal dealt a crippling blow to the Creek tribe, for Cow Tom it would have certain advantages. He had seen enough to know that whatever fate awaited his family in Indian Territory was preferable to the harsh bigotry of the white man's world. To Cow Tom the land of opportunity in 1837 was not America but a place beyond its reach: the land of the red and black.

The new Creek territory lay in an area which is now eastern Oklahoma. Each of the Southeast's so-called Five Civilized Tribes—the Creeks, Cherokees, Seminoles, Choctaws, and Chickasaws—was given its own section of Indian Territory and the right to govern it independently.

The tribes were also given the right to starve to death. The Indian Territory climate was hostile and unfamiliar to the newly arrived Creeks. They settled near the riverbanks along the juncture of the North and South Canadian rivers, only to find their homes washed away during violent storms. Warm sunny days turned suddenly to intensely cold nights. The soil was fertile but uncultivated, and there was a chronic shortage of tools.

Within a few years hard work and the communal treatment of property succeeded in making the undeveloped land habitable and preventing starvation. But the Creeks did not prosper in their new home. The tribe's population decreased steadily. Malaria and other unfamiliar diseases took their toll, as did the sickly spirits of those too depressed to adapt to a region so far from the land of their ancestors. Before leaving the Southeast there were more than twenty-four thousand Creeks. By 1859 only 13,500 remained.[18]

For thousands of Native Americans the forced removal west was a death sentence, the starting point of what they called "the road to disappearance." The misfortune of the Creeks, ironically, contributed to the prosperity of the tribe's blacks.

To the Creek slaves and their descendants migration brought with it an unprecedented opportunity to escape the influence of racist whites. During the same period in which the Indian population of the tribe was almost halved, the black population quadrupled.[19]

Yargee's slaves worked a five-day week. On weekends and evenings they managed their private gardens (they were expected to provide for themselves), conducted personal business, and attended a Baptist church he built for them. Unlike southern states, where slaves needed passes to leave their owner's property, most Creek slaves like Yargee's were able to come and go as they pleased. Discipline was so lax that slaveholders outside the Indian nation refused to buy them; auctioneers in neighboring states regularly complained of the extreme "difficulty of controlling them."[20]

Cow Tom and his family lived in Yargee's "Negro Quarters" (which resembled a small town), in a house clustered near those of other slaves for protection against hostile Plains Indians. Their log cabin probably had two rooms, dirt floors, and walls chinked with bright red earth and animal hair. They grew most of their own food, living on vegetables, rice, beef, pork, and wild game.[21]

Although the Creek nation maintained its political sovereignty for two generations, citizens found themselves interacting with whites far more in the West than they had in the Southeast. In addition to the missionaries, merchants, and schoolteachers, agents from the Indian Affairs department played increasingly important roles as Americans "tamed" the western frontier. Cow Tom's job as Yargee's interpreter expanded into a full-time career. As a member of the Creek Council, Yargee needed him for those meetings which included white officials. More important, he relied on Cow Tom as his business manager.

Yargee, among the tribe's wealthiest citizens, frequently found himself doing business with white merchants licensed to operate trading posts in the Indian nation. Like many full-blooded Creeks who spurned the white world, Yargee spoke no English, and would have been an easy mark for dishonest salesmen. Such recalcitrance became a license for blacks like Cow Tom to succeed: the more the Creeks depended upon their slaves to perform essential tasks, the more important these slaves became. Blacks were believed to possess an understanding of the white man's ways as well as his language, and most Creeks wanted nothing to do with either. Cow Tom negotiated many of Yargee's business deals, receiving bonuses, as was the tribe's custom, for successful transactions. If he handled the purchase of a dozen head of cattle he would often be given money, or a calf. In hopes of purchasing his freedom, Cow Tom saved every cent he could lay his hands on. Some Creek slaves were so adept at hoarding money that their Indian masters relied on them for cash loans.[22]

The unique status of black Creek interpreters like Cow Tom was documented by army colonel Ethan Allen Hitchcock, who kept a lively journal of his 1842 journey through Indian Territory. In each of his encounters with Creeks Hitchcock dealt through Negro intermediaries. When he met Yargee he reported that Cow Tom did all the talking for the wealthy chief, who barely knew a word of English. The interpreter used the opportunity to voice his own concerns. "Tom wants a school for his children," Hitchcock wrote. "Wishes he had a grist mill. Thinks the wheelwright employed by the government don't 'do well—seems like his work don't stand,' timber not seasoned."[23]

Within a few years of Hitchcock's report Cow Tom had accumulated enough money (probably about $400) to buy his way out of slavery. From that point on when he interpreted for Creek leaders like Yargee he was paid—and paid well. Because the U.S. government picked up the interpreter's tab during official negotiations, his was one of the highest-paying occupations in Indian Territory. The government paid black interpreters $2.50 a day—three times the salary of a blacksmith.[24]

As a free Creek Cow Tom was entitled to the same privileges as his red tribesmen. This meant he was allowed as much unclaimed land as he could manage. He selected a site along the Canadian River and quickly expanded his ranching business. Before long he was able to purchase freedom for his wife and daughters. By 1861 he owned seventy-five cows, sixty steers, one-hundred-ten hogs, eleven yoke oxen, ten horses, and five hundred bushels of corn.[25]

Cow Tom wasn't the only black Creek to prosper financially. Conditions within the tribe were so amenable to slaves that hundreds of others also managed to save enough money to buy their freedom. According to one U.S. government official, by 1861 an estimated 1,600 blacks—fully half those living in the Creek nation—were free members of the tribe.[26]

Most slaves who bought their freedom were part of the "Upper Creek" faction of the tribe, who lived apart from their Lower Creek tribesmen. Upper Creeks like Yargee made up the larger portion of the tribe and kept as far from the white man as possible. They rarely learned English or wore Western clothes. The "Lower Creeks," on the other hand, emulated whites, with whom they intermarried whenever possible. (Upper Creeks were more likely to be "full-bloods," or intermarried with blacks.)

The Lower Creeks were dominated by a handful of wealthy mixed-white families who saw themselves as southern aristocrats. Their children were sent to American colleges and their homes were crammed with furniture imported from the East. To build agricultural empires a few families each employed hundreds of slaves, whom they treated with the same severity as their white cousins in Georgia. Racial supremacy was important to them, and they were embarrassed by the racial mixing of their Upper Creek tribesmen. Because they were more interested in politics, better educated, and spoke some English, the aristocrats were able to control the Creek Council. In1861, when the Civil War arrived in Indian Territory, the Lower Creeks threw their support behind the proslavery Confederate states.

The secessionist government armed and trained supporters from the Five Civilized Tribes, who formed a fierce cavalry regiment and quickly took control of Indian Territory. The Upper Creeks supported the Union. Led by a highly respected chief named Oppotheyahola, they appealed to Washington, D.C., for the protection that had been promised them in long-standing treaties, but their request was ignored. When six thousand pro-Union Creeks and blacks tried to leave their homeland for Union-controlled Kansas, they were attacked and massacred by their Confederate brethren. More than fifteen hundred of them never made it. Those who survived huddled in squalid refugee camps in Kansas and waited for the war to end.[27]

Cow Tom never fled to Kansas. The Florida campaign against the Seminoles had given him a strong-enough distaste for the white man's wars to last a lifetime. Refusing to leave behind the ranch he worked so hard to build, he became a rare holdout in the Upper Creek community. The nearby towns were empty and his neighbors abandoned their homes when the Civil War reached the territory. But the seclusion of Cow Tom's farm, at first viewed as the misfortune of the freedmen for being the last to settle in the area, became an important advantage in a civil war. Few people knew he was there, and his family was able to hide from the occasional Confederate patrols which stumbled upon his ranch.

Were it not for the Battle of Honey Springs, the bloodiest fought in Indian Territory, Cow Tom might have been able to hide out for the entire war. The battle, in early 1863, was a decisive victory for the Union. It was one of the first Civil War engagements in which African-American soldiers played a determinative role in battle, and it marked the turning point in the war for Indian Territory.[28] It made refugees of hundreds of Confederate soldiers, who escaped into the nearby countryside. Like locusts the deserters occupied whatever homes they came across and took what they needed. A few months after the battle hundreds of armed men moved into Cow Tom's area along the Canadian River. By the time they reached his house Cow Tom and his family were gone. Taking only the clothes on their back and provisions for the road they disappeared into the tall hay fields.[29]

After a week of nightly travel Cow Tom reached the protection of Fort Gibson, located in the Cherokee nation near the Creeks' northern border. The Union outpost, although reoccupied just four months earlier, was already packed with thousands of Cherokee, Creek, and black refugees. Their tents and cooking fires dotted the sparse landscape, extending for miles beyond the high stockade walls of the fort.

The federal authorities at Fort Gibson were most concerned with the largest segment of the camp population, some six thousand Cherokees. About one thousand Creeks and hundreds of black refugees fell between the government's bureaucratic cracks. The tribe's few established leaders were in Kansas, where the official relief for the Creeks was being handled. Those who had found their way to Fort Gibson were leaderless, and few spoke any English. The tribe's blacks were even worse off. All their army-age men had enlisted (Cow Tom, already past fifty, was ten years past the cutoff age for recruits), leaving the women, children, and elderly at the mercy of federal officials. Neither group was able to draw rations. Some were reduced to tagging behind the mules of the military's supply caravans, scooping up bits of corn that fell to the ground.[30]

Soon after arriving at the fort Cow Tom took charge of the refugee Creeks. There were no other chiefs or tribal councils to authorize his role. He assumed leadership in the manner through which other Creek chiefs were occasionally elevated: by a combination of capability and necessity. Cow Tom was bilingual, intelligent, and accustomed to negotiating with white officials. Somebody had to interpret and speak for the Creeks, and with a regiment of black soldiers helping to protect Fort Gibson it would have been hard to construe a racial argument for preventing the most qualified man in the camp from doing so. Leadership roles in any group are rarely handed over voluntarily to those of a different class from existing leaders. Cow Tom was not made a chief by the Creek hierarchy. He *became* a chief—and was probably the first full-blooded Negro to assume the role of an Indian chief in the history of North America.[31]

Most of the pro-Union Creek refugees were stuck in horrible conditions at other camps in Kansas. They also depended on a black Creek interpreter to communicate their needs to the federal government. The job fell to Harry Island, a friend of Cow Tom's who, during the 1850s, had fathered a child by Cow Tom's daughter Matilda. Between the two men virtually all the tribe's important negotiations took place. The nature of their association was entirely unknown to their red brothers at the time—or Oklahoma historians later.[32]

Although they often worked independently Cow Tom and Harry Island dominated the tribe's interpreter market, learning more about the tribe and its relations with the government than even the highest-ranking Upper Creek chiefs. They were looked to for guidance as well as interpreting. The war made mortal enemies of their better-educated Lower Creek countrymen, upon whom Upper Creek leaders had always relied to manage tribal politics. The war had delivered the loyalists into a state of helplessness, blurring the distinctions between master and slave, black and red. They were the Creek people, chased from the same land, sometimes born of the same father. Both had been stripped of their possessions, their homes, their communities. The traditionalist Upper Creeks who had spurned white society were now dependent upon its supply wagons for their food. Blue-uniformed men whose language they could not understand controlled their fate. Between the Indians and the incomprehensible stood one figure: the black interpreter. Cow Tom and Harry Island were willing to communicate the tribe's needs to the government, but they would communicate for the whole tribe—black as well as red.

Island's input was quickly apparent. While working for Principal Chief Sands (who succeeded Oppotheyahola), he saw to it that the Creeks accepted Lincoln's 1863 Emancipation Proclamation with an unusual flourish, recognizing the "necessity, justice, and humanity" of an end to slavery. He also secured a commitment allowing former slaves tribal land once the war was over. Resettlement, however, proved far from easy.[33]

In May 1864 the army accompanied nearly five thousand Creek refugees from Kansas to Fort Gibson, bringing the population around the fort to nearly twenty thousand and creating one of the largest cities in the Southwest. It was like an overnight Gold Rush town, only there was no gold. There was barely any food, either, and little anyone could do about it.

Chief Sands took charge of the tribe from Cow Tom, delegating to him chiefdom of the Creek freedmen. Harry Island continued as the official U.S. interpreter. When Sands wanted to speak to government officials or officers he would speak through Island. And when the blacks were to be consulted for decisions which affected them, Island was sure that Cow Tom was at least as well informed as the head of the tribe.[34]

When peace finally came to the area on May 27, 1865, the Creeks repopulated a land devastated by war. A Mafia-like ring of organized cattle thieves from the Union army had stripped their nation of its greatest resource—some 300,000 head of cattle. Warring factions had also destroyed homes and burned farmland. Thousands had been killed by war, famine, disease, and exposure. Between 1859 and 1867, one quarter of the tribe perished.[35]

It was in this state of destitution that the Creeks were forced to negotiate a new treaty with the government. The United States took full advantage of the situation. The fact that a majority of the Creek, Cherokee, and Seminole tribes had risked their lives out of loyalty to the United States was immaterial. Federal law stipulated that if any faction of a tribe signed a treaty with the Confederacy, then all existing U.S. agreements with that entire tribe were invalid. By making the loyalists suffer for the sins of their secessionist tribesmen, government officials were able to seize major portions of land for use by other displaced Native Americans and, more important, powerful railroad interests.[36]

In September 1865 leaders of the five tribes of Indian Territory were summoned to Fort Smith, Arkansas, for a peace council with America's top Indian officials. The Creeks brought the most delegates to the conference; their twenty-two representatives included both Upper and Lower Creeks, as well as five black delegates, among them Cow Tom and the official interpreter, Harry Island.

The most significant evidence that Cow Tom and Island had secured a unique role for their tribe's blacks was that none of the other tribal delegations called to Fort Smith included even a single black tribesmen. The Creeks problack slant was bitterly resisted by the Lower Creek's Confederate delegation. They demanded compensation for their liberated slaves and objected strongly to a government demand that each tribe provide for its former slaves. Soon after arriving the prosouthern faction proclaimed, "We agree to the emancipation of the Negroes in our nation but cannot agree to incorporate them upon principles of equality as citizens thereof—and we cannot believe the government desires us to do more than it has seen fit thus far to do . . ."[37]

The Confederate Creeks, led by their well-educated chief Samuel Checote, had a strong argument. At this stage in history, the U.S. had done nothing to facilitate the enfranchisement of blacks into American society.[38] President Andrew Johnson, Lincoln's successor, left the status of southern freedmen to the individual ex-Confederate states. They adopted "Black Codes" prohibiting African-Americans from voting, holding office, assembling, or bearing arms. Many states forbade them to leave their jobs, and allowed police to assign them as laborers for chain gangs. Carl Schurz, one of the organizers of the Republican Party, called the Black Codes "a striking embodiment of the idea that although the former owner has lost his individual right of property in the former slave, the blacks at large belong to the whites at large."[39]

Cow Tom was determined to cut a better deal for ex-slaves in the Creek nation. He and the other delegates convinced the Upper Creek leaders to support black enfranchisement in no uncertain terms. On the final day of negotiations Harry Island read a statement on behalf of the loyalist Creeks. "We are willing," it said, "to provide for the abolishing of slavery and settlement of the blacks who were among us at the breaking out of rebellion, as slaves or otherwise, as citizens entitled to all the rights and privileges that we are."[40]

The government had a harder time convincing leaders of the Indian Territory's other tribes to accept their former slaves. Officials tried to convince leaders of the Five Civilized Tribes that by incorporating the territory's ten thousand blacks they would "rapidly augment their numbers and power."[41] Most Confederate Indians did not buy it. The Chickasaw and Choctaw tribes blamed the blacks for the South's defeat and initiated a reign of terror, regularly murdering those who passed through their country.

Black Creeks were a special case. According to a federal official's field report soon after the war they were accepted as equals and often became "the most industrious, economical, and, in many respects, the most intelligent portion of the population of the Indian Territory."[42]

Although in the minority, the Confederate Creeks continued to resist allowing the freedmen into the tribe. They were motivated by financial concerns as well as racism. Blacks officially represented nearly two thousand of the roughly twelve thousand citizens of the Creek nation. Granting them rights to tribal funds would dilute the average Indian's stake in government-annuity disbursements by more than 15 percent. (The payment was like interest on a debt from federal purchases of tribal land.)

In early 1866, when Checote heard that an official delegation of loyalist Creeks had been called to Washington to make final the postwar treaty, he quickly dispatched a Lower Creek delegation to the Capital to argue against integrating the tribe. The southern delegation, although not officially recognized by the government, argued that the treaty would not be valid unless both factions of the tribe signed it. Unlike the Upper Creek leaders (who signed treaties with an X), the Lower Creeks were able to write—and write eloquently. In a March 18, 1866, letter to federal officials claiming to represent a majority of the tribe, they argued:

> . . . We are determined to protect the interests of the Negroes who were among us before the war. . . . But we believe that our ancient care and kindness ought to be a sure guaranty that their interests and welfare will be safe in our municipal jurisdiction. We never have, and have not now, any disposition to injure or tyrannize over them. Still, we can never recognize them as our equals. We are honest and candid in this declaration. . . . It is, we conceive, contrary to nature and nature's laws. The antipathies of race among Indians are strong, if not stronger than they are among the whites. The government of the United States is very strong; we are very weak; it can, and may, force us to things repugnant to our nature; but it cannot change our honest convictions and faith any more than it can change the skin of the Ethiopian or the spots of the leopard.[43]

The southern delegation were better diplomats than the Upper Creeks. Historians have noted that the whole tribe would have come out far stronger had the Lower Creeks gotten their way[44]—the whole tribe, that is, except for the blacks.

After six months of negotiating a treaty acceptable to both factions the Lower Creeks refused to sign unless the loyalists relented on the adoption of freedmen. By this time federal officials just wanted to get the agreement signed and complete their land grab. They suggested that Chief Sands and his people compromise on the freedmen issue. But Harry Island, the Upper Creek's interpreter, prevailed upon him to stand strong. Commissioner of Indian Affairs D. N. Cooley's 1866 *Annual Report* described how the conflict was resolved:

> It appeared at one time as if all negotiations must fail, and the Commissioners, knowing the necessity of some settlement . . . were disposed to urge the national delegates to yield the point for the present, but they held out firmly for the freedmen, urging that when the brave old Opotheyohola, resisting all the blandishments of the rebel emissaries, and of his Indian friends, stood out for the government, and led a large number of his people out of the country, fighting as they went, abandoning their homes, they promised their slaves that if they would also remain faithful to the government they should be free as themselves. Under these circumstances the delegates refused to yield, but insisted that that sacred pledge should be fulfilled, declaring that they would sooner go home and fight and suffer again with their faithful friends than abandon the point. They were successful at last, and the treaty guaranteed to their freedmen full equality.[45]

Signed by both the loyal and Confederate delegates on June 14, 1866, the freedmen's clause in the treaty read: "And inasmuch as there are among the Creeks many persons of African descent, who have no interest in the soil, it is stipulated that hereafter these persons residing in said Creek country . . . and their descendants . . . shall have and enjoy all the rights and privileges of native citizens, including an equal interest in the soil and national funds, and the laws of the said nation shall be equally binding upon and give equal protection to all such persons, and all others, of whatever race or color, who may be adopted as citizens of members of said tribe."[46]

With the stroke of a pen the Creek chiefs had agreed to do more for their former slaves than any other group in America. To what extent their action on behalf of the freedmen can be attributed to the men who interpreted and spoke for them is still a matter of historical dispute.

In an unpublished dissertation on blacks and Creeks written in 1937 at the University of Oklahoma, Sigmund Sameth noted a widely accepted viewpoint. "A present-day Indian informant," he reported, "says, 'The reason the niggers got equalization was because they had a nigger interpreter who looked out for his own people.' "[47]

This belief, somewhat exaggerated, continues today through oral history. Erma Threat, now in her mid-sixties, recalls that her grandfather, Jake Simmons, Sr., told her of the interpreter's role in the negotiations. "The Indians," she says, "didn't want the black to have the 160 acres of land. But when he [Cow Tom] would interpret to the white man, he would interpret that the Indian had said, 'Let the blacks have the land.' "[48]

Despite such conjecture it is likely that the Creek chiefs had a fair sense of what they were doing. Cow Tom and Island had pressured them, over the course of a three-year campaign, into making public commitments to adoption. Afterward, the black leaders used their position whenever necessary to remind them that they were morally bound to abide by such commitments. And unlike the U.S. government, once the Creeks gave their word they stuck to it. To the loyalist chiefs the assurances given to Cow Tom and Harry Island had become a matter of honor, and in their honor-bound world this was far more important than the dilution of tribal annuities.

Among the other tribes in Indian Territory only the tiny Seminole nation (numbering fewer than two thousand, including blacks) accepted the full enfranchisement of their former slaves. The Cherokees disputed elements of adoption for years, and the Chickasaws and Choctaws never agreed to it.[49]

As long as it didn't cost whites anything American officials were absolutely determined to see justice done for the black Indian. The United States took nearly a century to grant the millions of freedmen in its own borders the same rights as Creek freedmen enjoyed after 1866. On an economic level adoption into the tribe's communal land trust proved far more beneficial to blacks than any program ever devised by the federal government—even in modern times.

Once granted land the Creek freedmen prospered. By the end of 1866 the federal Indian agent noted that blacks "are today further from want than are their former masters."[50]

As patriarch of a large family Cow Tom spent most of his time building his private estate, leaving tribal politics to a younger generation of black Creek leaders. Citizens of the Creek nation were entitled to all the land they could put under fence and cultivate. Cow Tom staked out a small plantation thirty miles southwest of Muskogee, near what is now the town of Boynton, Oklahoma. Several laborers and their families lived on the farm, along with Cow Tom's two daughters and their children. So successful was his farm that when a description of Indian Territory was published in the 1870 *Annual Report of the Commissioner of Indian Affairs,* Cow Tom was singled out and described as a superlative example of a prosperous Creek farmer.[51]

A traditionalist, Cow Tom preferred self-sufficiency to cash crops. His family pounded its own corn without a mill, crushing it in a mortar with a wooden pestle. They had no cotton gin, and sifted and carded their cotton by hand to make all their clothes. Dyes of pink and red came from the sycamore and red oak trees; sumac and copperas made tan colors. He used old-fashioned farming tools and had no use for modern medicine. The nearest doctor was in Fort Gibson, thirty-three miles away. Like their Indian neighbors Cow Tom's family relied on herbal remedies and magic to solve those problems they believed were meant to be solved.

Cow Tom hunted and raised a few chickens but devoted most of his energy and resources to building a cattle business. He eventually handled as many as a thousand head a year.[52]

The Confederate faction of the Creek tribe continued to resist the freedmen's adoption into the tribe. In November 1866, when the time came to distribute $200,000 from the forced sale of Creek land, Samuel Checote initiated a campaign to prevent blacks from being counted in the tribal role of those eligible for per-capita disbursements. He went over the head of his tribe's Indian agent, who wanted to include the blacks, and appealed to the southern superintendent of Indian affairs. He argued that the Treaty of 1866 guaranteed blacks a right to share in school funds, not this special postwar disbursement.[53]

The superintendent deferred the decision to a new commissioner of Indian affairs in Washington. It was a political hot potato in an ever-expanding stew of discontent over reconstruction. Congress was trying to force a recalcitrant Andrew Johnson to accept civil rights legislation. The President responded by twice vetoing a Freedmen's Bureau Bill and fighting against the Fourteenth Amendment, which extended citizenship to blacks.

Not surprisingly, when Johnson appointed a new Indian affairs commissioner it was a man whose sympathies in the Creek tribal dispute were clearly in line with his administration's antiblack enfranchisement stand. Early in the morning of March 14, 1867, the very day a local agent was to begin distributing $17.34 in gold to every member of the Creek tribe, regardless of race, a horseman arrived with instructions from Washington to remove the freedmen from the list.[54]

The decision indicated a shift in tribal power. The loyalists' honeymoon with the federal government had lasted less than two years. From this point on the superior education and racial affinity of the part-white southern Creeks would once again allow them to dominate tribal politics. By controlling the machinery of democracy they gained full control of the tribal government in a hotly disputed election.[55] Once in power they set their sights on the unusual status of their tribe's blacks.

The ex-slaves were troubled. Although the tribe ostensibly recognized their rights and provided black representation in both chambers of its congress (the "House of Lords" and "House of Warriors"), the black Creeks worried about their future. They had been excluded from an official census, a record from which future allotments and disbursements might depend. It was almost as though they had been written out of the tribe, that the guarantees secured by Cow Tom and Harry Island had been wiped out. What next might Checote's party decide the freedmen weren't entitled to? School funds? Farming implements? Land allotments?

Cow Tom and Island reentered politics. With their Upper Creek allies out of power they had no choice but to act on their own. With other black Creeks they appealed in writing to the boss of the commissioner of Indian affairs, Secretary of the Interior O. H. Browning. Browning, another Johnson appointee, considered the case closed.[56] They decided that their only hope was to take the case beyond the Indian administrators to the most powerful institution in the land, the American Congress.

It was a bold ploy. Checote had gone over the Creek agent's head to a more sympathetic commissioner. They would go over the commissioner's head to the Untied States Senate.

Cow Tom, Harry Island, and another interpreter named Ketch Barnett traveled by wagon to Lawrence, Kansas, where they boarded a train for Washington. Their timing could not have been better. A year earlier their cause would not have stood a chance. As it was they arrived in the spring of 1868. "Radical Republican" Reconstructionists had finally been able to muster a two-thirds majority in Congress, which they used to overturn President Johnson's vetoes and pass civil rights legislation. They divided the South into five military districts and used the army to ensure that blacks were awarded full political representation. Theirs was a vision of an egalitarian nation, an America for all the people.

Cow Tom appealed to Congress at the height of the radical Reconstructionists' power, right after Johnson was nearly impeached by the Senate on May 16, 1868 (he survived by just one vote). His delegation appeared before the Senate's Committee on Indian Affairs, where their charges that Presidential appointees assisted the Confederate Creeks in depriving them of promised funds added timely fuel to the anti-Johnson fire.[57]

On May 27, 1868, Kansas senator Samuel C. Pomeroy brought a resolution before the Senate: "That the Secretary of the Interior be directed to inform the Senate, at as early a date as possible, the reasons why a large number of persons registered as Creek Indians by the Creek Agent in the spring of 1867 were stricken from said rolls and payment of their per capita dividend refused." It passed unanimously.[58]

Johnson's Secretary of the Interior denied that blacks were entitled to the funds. The committee reported back to the full Senate that President Johnson's appointees were unwilling to carry out important treaty stipulations. But they, the radical Republicans, had the congressional clout to force federal officials to adhere to the law. Johnson could harass their reconstruction initiatives but not block them. Later that summer, Congress, with expressions of outrage at the behavior of the "rebel Indians," set aside $30,882.54 of the tribe's annuities for the freedmen.[59]

A law is not truly a law until it has been tested and proven effective. Had Cow Tom, Island, and Barnett not challenged their own chiefs and the hierarchy of the government's Indian Affairs establishment, the adoption of the blacks in the Creek Treaty of 1866 might not have been worth the parchment it was written on. The value of the delegation's actions is best characterized by something Frederick Douglass, America's most prominent black abolitionist, said a decade earlier: "Power concedes nothing without demand. It never did and it never will. Find out just what any people will quietly submit to and you have found the exact measure of injustice and wrong which will be imposed upon them, and these will continue till they are resisted with either words or blows, or with both."[60]

The $17.34 per-capita payments, made the following summer, were not what mattered—although for a large family the payment almost equaled the value of a season's crop. The important thing was the willingness of the black Creeks to stand up for their rights, and their ability to ensure that those rights were respected. The Creek freedmen were never again left out of the tribe's disbursements. When tribal land was broken up into individual plots three decades later this precedent would provide the blacks of the Creek nation with an entitlement unequaled in the American experience.

The trip to Washington was Cow Tom's last public act. He returned to the ranching life he preferred, raising cattle and the swarm of kids who looked upon him as a father. He died on June 1, 1874, at the age of sixty-five. His wife, Amy, died two years later. The couple are buried a few feet from each other at the Cain Creek cemetery, near the center of the Creek nation. The small, overgrown cemetery borders an old white church on a dirt road. It also holds the remains of Cow Tom's children and grandchildren, some of whom chose to be near the obscure resting place of their family's chief nearly a century after he died.

It took more than a hundred years, until Jake Simmons, Jr., was critically ill, for the Creek tribe to undo what Cow Tom had achieved. In 1979 the Creek leaders rewrote their constitution, excluding freedmen and intermarried couples, as well as their descendants, from the tribal roll.[61]

The road of opportunity paved by Cow Tom stretched well beyond his death. A signal was sent to those who would repress the progress of the tribe's black citizens, a signal so clear and strong that it lasted right through to the next century.

1

CHILD OF THE FRONTIER

A man doesn't want to have to milk the cows at four-thirty every morning. A man doesn't like being told what to do all the time.
—Jake Simmons, Jr., at the age of five

Jake Simmons, Jr., was born on January 17, 1901. His forty-six-year-old mother, Rose, on her ninth child, needed no help with the delivery. She closed herself alone in the bedroom. As in the tradition of many Native American tribes she didn't lie on a bed but squatted, half-standing, letting gravity draw the baby out. By this time Rose was experienced at self-deliveries. (Once, when she unexpectedly went into labor in a hay field an hour from the house, she managed outdoors.) She used sharp, sterilized scissors to cut perfectly the umbilical cord (none of her children ended up with a ruptured, or even protruding navel). A few hours later, Rose emerged from the bedroom, a crying infant in her arms. She was ready to get on with the day's work.[1]

Rose Jefferson was Cow Tom's granddaughter. When she married Jake Simmons, Sr., in 1884 she was not what many men of the late nineteenth century would have thought a prime catch. Instead of being large and voluptuous she was tiny and trim. At nearly twenty-nine she would have been considered an old maid for a childless woman—and her face bore a coarse, hardened look. While other eligible wives were obedient, she was introspective and authoritative. Like her grandfather she had dark black skin, and like her grandfather she was more of a traditional Creek than an American. She never tried to alter her hair to the styles of the times but kept it cropped short in a "natural cornrow," generally wrapped with a scarf. Her legs were always covered with long dresses, invariably in dark colors. She had little use for socializing or small talk, and didn't think of herself as particularly attractive. Her only store-bought clothes were pointy black peg-tied shoes, fashionable for conservative women of her day. Female relatives recall that her feet were of extraordinary beauty—and the only physical characteristic she was proud of.[2]

Born about 1855, Rose grew up on Cow Tom's farm. The old chief was the central figure of authority in her family. Rose's mother, Melinda, married a black Creek politician who, in accordance with Creek tradition, moved in with his wife's family after they wed. Although her family eventually inherited part of Cow Tom's large estate, Rose was far from spoiled. As the oldest girl she looked after her four brothers and sisters and took them to an integrated school every day. When not in class she was expected to help feed the relatives and ranch hands who formed the settlement's large extended family. Although unmarried, Rose was determined to become independent. She left her family for the tribe's busy capital town of Okmulgee, where she used her culinary skills to land a cook's job with a wealthy Indian family.[3]

With few social skills and her youth beyond her, Rose might have spent the rest of her life working in someone else's kitchen had Jake Simmons not heard about her from a group of Indian ranchers visiting his employer. After listening to them talk about the unmarried black woman who had cooked a household of visitors a sumptuous feast in Okmulgee, nineteen-year-old Simmons decided to pay her a visit.

Jake was impressed by what he found: a sharp-witted, forceful dark-skinned woman ten years his senior. Their son, John Simmons, recalls his father telling him, "I started courting her the same night I met her." For the next few months Jake visited Okmulgee every chance he got. It wasn't a passionate affair but it made a lot of sense, which meant more to Simmons than anything else. In a twenty-page biographical sketch dictated to a grand-niece late in life he recalled, "I thought I was a man. I got married while my boss man was away on Saturday night, July 24, 1884, to John Jefferson's daughter by the name of Rose Jefferson[;] her father at that time was the wealthy colored man in this part of the country, but I kept on working and I didn't see my wife for two months [although] I was only twelve miles from her."[4]

Jake was an impulsive boy with a clear idea of what he wanted in a wife: the domestic stability he had never known. His own mother had been unable to provide him with a traditional home. She was a half-Indian, half-black Creek named Lucy Perryman. Lucy was seventeen when Jake was born on September 22, 1865, in the Confederate refugee camp of Boggy Depot. The identity of his father is difficult to ascertain. During his life he told people that his father was one of two men, either a white man or a Creek rancher for whom he later worked.[5]

The white man was Jim Simmons, who came to Indian Territory from Missouri in 1851, and was "adopted" into the Cherokee tribe. At the outbreak of the Civil War he enlisted in the Confederate Army. For a short time he was thought to have settled with Lucy and the boy near the Creek town of Tullahassee, where he operated a butcher shop. He tried to negate his wife's African heritage by forbidding her to fraternize with black relatives. When she refused he abandoned both her and five-year-old Jake, leaving them destitute.

Men were scarce after the bloody war; Lucy's own mother was also single. Lucy moved from relative to relative, often leaving Jake for a month or a year with whoever was willing to care for him. When the boy was eight he was passed off permanently, this time to his grandmother.

Jake's grandmother had no livestock or horses; she was dirt-poor. They moved three times in the next two years, always one step ahead of starvation. Finally settling outside the town of Haskell, they were able to beg food from a nearby family until they brought in a small crop.

The boy grew up very quickly. Although he occasionally attended school it held little interest for him; he would go for a few months, then drop out to earn money. Survival was a far more pressing matter than arithmetic. He learned to get by, to ask for odd jobs in return for food, to spin a convincing tale when it suited his needs. Most of all he learned how to work hard and be anything to anybody. Long-faced and lanky, with big ears and the racial characteristics of a very light-skinned black Creek, he was one of hundreds of freewheeling frontier cowboys of indeterminate background who would help rebuild Indian Territory after the devastating Civil War.

To support his grandmother Jake worked on a neighbor's farm, sold pecans, fished, and cut logs for a new railroad line. He also got a job making and delivering sorghum molasses, but that was short-lived. After begging a pan of cornbread from a customer he ate so much of the sweet amber syrup that he took horribly ill and was fired. At the age of fourteen he decided to strike out on his own. He did his best to leave his grandmother in a secure state, having earned a hog and two bushels of corn by splitting fence rails. He then prepared to "run off." A day of legwork would take him sixteen miles and bring Simmons into a new world. In his short oral history, transcribed by his grandniece Pearl, he told of the day he left home:

> I got up early that morning. I made my fire and crept to the bed to see if my grandmother was sleep then I lit out. I made to my Uncle Dick Harrison two mile west of where we were living and got my breakfast and he ask me where I was going I told him I were running away from grandmother to hunt me a job. He says where are you going I told him I heard Bluffer Miller wanted a boy I'm trying to make it there he say I would carry you [on his horse] but your grandmother would give me a rackin about it so you have to go best you could so I left. . . . I walk on through Sever's wild cattles with a stick in my hand I made it to Sever's ranch about 11 o'clock. Mrs. Severs ask me who I was I tried to pick out the biggest man in the family, I says William McIntosh [actually Jake's great-uncle] so she ask me where I was going I say to look for a job I told her I heard Bluffer Miller wanted a boy I was trying to make it there so she say why that my brother-in-law you stay and get your dinner because you have to walk over six miles and she gave me a chair. . . . I struck out again got down to the rock creek close to the [Miller] house and my shoes broke in two I had on so I had a piece of knife and I got some hickory bark and tie it up. . . . he [Bluford Miller] and his wife were sitting on the porch. . . . I ask him for a drink of water, he say yes just go around to the well and help yourself, he gave

me cup then I asked him was his name Bluffer Miller he says yes, I says I'm looking for a job McIntosh told me you wanted a boy. . . . he say well I told Mike that but I don't need none now, so I says do you know where I can get a job he say Charley Clinton might could use you he live up there at Mounds, and I say how far is it, and he say 16 miles, and I ask him could I make it there this evening he says no it too far, so he say you could stay all nite, and I'll carry you in the morning, so he says where you from and who are your people, I say William McIntosh. . . . he say you make yourself welcome me and my wife going to work in the garden, you walk a long way, so I say I'm not tired I'll help you. . . .

Jake helped Miller all evening. The following morning it rained, so Miller told him they would have to delay their trip until the weather broke. Meanwhile there was wood to cut for rails. Every day he gave the teenager a different job. By the end of the week taking him to Charley Clinton was the last thing on Miller's mind. He drove Jake to town, bought him new shoes and underwear, then offered him a job for six dollars a month.

Simmons couldn't have fallen in with a better employee. Bluford Miller was one of the most successful men in the postwar Creek nation. He had a reputation for integrity and his ranch was near the region's most important north-south cattle trail. His estate eventually held enormous herds of cattle and covered tens of thousands of acres.

Miller, an orphan himself, liked nothing more than giving opportunities to poor, hard-working young men of all races. Like a one-man YMCA he helped raise Jake and countless others, many of them orphans. He took a special liking to Simmons, who arrived before he had any children of his own. Patient and instructive, he taught the boy riding, wagon driving, shooting, and cattle care. Miller found Simmons to be an eager learner. Before long the rancher was entrusting Jake with thousands of dollars to make major cattle purchases and treating him like his own son.[6]

Some members of the Simmons family believe that Jake was indeed Bluford Miller's son, and that it wasn't blind fate which brought him to the rancher's doorstep. Jake's son, John Simmons, born in 1899, is the last surviving member of his generation. He says that his father told him that Bluford Miller was his real father. He admits that other relatives, including his own brother Jake Simmons,

Jr., disputed this. It might not have been true, or, in light of the prominence of the Miller family, it might simply have been a difficult thing to acknowledge. When interviewed in 1937 for an exhaustive WPA project called the Indian Pioneer History, Jake told the interviewer that his father was Jim Simmons, the Missouri white man.

As Miller's ranch grew so did Jake's responsibilities, especially trading cattle. Beef was a booming business. The Creek nation's bluestem, sage, and buffalo grass grew higher than a man's head, making the area a cattleman's paradise. Located between the plentiful herds of Texas and the major rail yards of Kansas, Creek Country became a conduit for much of the great northern cattle drive. Hundreds of thousands of head were brought by railroad from Texas, fattened on ranges in Oklahoma for summer and fall, then sold to stockyards in the east.

When he was not buying or selling, Jake was a "cow puncher," living in the saddle for weeks during biannual roundups across the immense prairies. In 1937, nearly half a century later, he fondly described his fellow cowboys for a WPA interviewer: "The range hands, all of them, were jolly good fellows. . . . They loved to play pranks on each other and, not because I am a cow puncher myself, I am compelled to say that they were brave men, hated a thief and a coward and despised lawlessness in any form."[7]

Lawlessness in many forms was prevalent in the postwar Indian Territory. Because it lay outside the U.S. border, the area was seen by outlaws of all races as fair game for robbery, livestock theft, and the evasion of authority. A mounted tribal police force called the Creek Lighthorse did its best to keep the peace. In Muskogee most of the officers were black.

The federal government also employed many blacks to work the Indian Territory beat. Although each tribe had its own legal system, only soldiers and U.S. marshals could arrest Americans—then bring them back to "Hanging Judge Parker" in Fort Smith for trial. Bass Reeves, the most famous of the area's marshals, was a large, fearless black man who always got his man. During his thirty-two-year career he killed fourteen men. A steel breastplate and a lifetime of good luck allowed him to retire without ever being wounded, despite regular gunfights in which bullets cut his belt in two, knocked buttons off his coat, split his hat brim, and sliced the bridle reins held in his hand.

The most impressive display of black power in Indian Territory was the presence of the Black Cavalry, which protected the area's tribes from hostile enemies. Created by a special act of Congress in 1866, the Ninth and Tenth Cavalry comprised 20 percent of all mounted troops in the West. Native Americans were so impressed by their horsemanship, ferocity, and appearance that they dubbed the cavalrymen "Buffalo Soldiers," after the animal they revered. Despite constant discrimination from both the army establishment and white soldiers, the Buffalo Soldiers won a dozen Congressional Medals of Honor and stuck to the cavalry with far greater enthusiasm and commitment than their white counterparts. In 1876, the Ninth Cavalry had six desertions, and the Tenth had eighteen, while Custer's Seventh had seventy-two, and the Fifth Cavalry had two-hundred-twenty-four.[8]

During the two extensive interviews Jake Simmons, Sr., gave late in his life he made no mention of whether his own six-gun ever saw action during the violent age of his youth. His last surviving son, John ("Uncle Johnny" to family and friends), says that it did. John keeps his father's Colt .45 revolver, patented in 1871, tucked under a couch in his Oklahoma City home. He was eager to show it to me during a visit—he is known to fire it annually on New Year's Eve. "See the notches on it?" he said, his eyes brightening as he showed off his most cherished possession. "You can tell just how many men he killed with that gun who disobeyed the rules. There are seven notches on it. You see 'em?"

John says that his father was once a deputy in the tribe's lighthorse police. As there were no prisons in the Creek nation, deputies were often required to chain prisoners outside their homes until they could be tried.[9] John claims that most of his father's victims died trying to escape. "In the summertime you kept him chained to the tree like he was a cow," he explains. "You feed him and carry him to the bathroom and carry him water. A lot of them would accept this tradition [but sometimes the prisoner would], get out and get rowdy, want to kill him. It was eye for eye, tooth for tooth then."

Jake Simmons never thought of pursuing a career in law enforcement. He wanted something more stable for his life, something more traditional. He wanted to run a successful family ranch like Bluford Miller. And he wanted to marry the type of woman who would best help fulfill this dream. He had little doubt that Rose Jefferson was such a woman.

Throughout his life few who did not know Jake Simmons well would have identified him as a Negro. His complexion was identical to that of many Indian Creeks, and he had fair hair and blue eyes. Had he wanted, Simmons could have easily "passed," married an Indian woman, and raised light-skinned children. Bluford Miller was eager to help him choose an Indian bride. Black men at the time were regarded by many Native Americans as better farmers and husbands, and a well-recommended man of Simmons's complexion could probably have had his choice of many Creek women.

As a young man John Simmons sometimes wondered why his father selected the bride he did. When he was old enough he asked him, "Dad, why did you marry a black woman when you were reared among the Indians and the whites?"

"Your mother had sense," his father replied. "My people [Indians] were lazy and didn't have any sense at all—especially the women." Rose's education was an important consideration. "Dad couldn't read and write," John reflects, "and my mother learned him to read and write. Now what would keep you from marrying a woman like that?"

Donald Simmons, Jake Sr.'s grandson, remembers another factor: his grandfather was impressed by Rose's distinguished ancestry. He told Donald that he married Rose because she was nearly "all African" and "her family had no history of slavery—he didn't want his children raised with those kinds of feelings, that they were less of human beings than anyone else." Also, Simmons is quick to add, "She had a lot of money."

Erma Threat, a granddaughter of the Simmonses, thinks Rose's money—and her ability to manage it—was more than an afterthought. "He thought Indians had no sense, and that the kind of white woman he could marry didn't have anything [financially]. My grandmother was smart, she was thrifty, and she had something. Neither of them were affectionate and I don't think they knew too much about love. But he had an eye for business. His marriage was three-quarters business and one-quarter love."

Rose's jet-black complexion also furthered Jake's interest in marrying her. Ophelia Brown, Erma's cousin, says that her grandfather "didn't like light-skinned ladies." When the family would be sitting around, she remembers Jake saying, "I always said that when I get married I'm gonna marry me a black-skinned lady." In public he made an even bigger point of his pride in Rose's skin color. While engaged in conversation at a crowded picnic or church supper he'd pause, then announce, "I may have white skin, but my wife, Rose, is black as tar, and I'm proud of it. Where are you, Rose? Stand up so everyone can see how beautiful your black skin is!"[10]

Bluford Miller treated his newly married hand much like a traditional son. He brought a friend all the way from Kansas to teach Jake how to set up his own farm, then gave him a $280 interest-free loan to construct his own twelve-by-fourteen-foot log house. The house was built in a fertile area nourished by surrounding creeks. The newlyweds would remain at the same location for more than fifty years, some twenty miles west of Muskogee and four miles from the current town of Haskell, Oklahoma.

Rose and Jake's early years together were rough. Their first crop was nearly destroyed by a hailstorm. He worked at odd jobs like rail-making to earn cash, sometimes leaving his wife alone to tend the cotton, corn, and children. He used the money to buy forty head of cattle and two horse teams, then hired his own hands to help cultivate fifty-five acres of corn and ten of cotton.

Simmons shipped small herds of cattle to buyers in St. Louis every season. Eager to expand his business, early on in his career he and a couple of partners borrowed $7,000 from a Missouri financing company to buy 315 head of Arkansas cattle. Prices plunged the next year, and they were forced to sell at a loss. Instead of making a killing Simmons was nearly wiped out. It took a $600 loan from a local bank to cover his share of the shortfall.[11]

A cattleman's life involved periodic jaunts away from home. Simmons had to travel across the region to buy, move, and sell his herds. He enjoyed the opportunity to see other places, and also enjoyed the opportunity to see other women.

Like many cowboys of his day, Jake Simmons considered himself a ladies' man. Although he had big ears and a roughly weathered face he was persuasive, and easy to be with. He also was not averse to flashing the large wads of cash he carried after cattle sales. According to John Simmons, the rancher's favorite come-on was "If you got the time, honey, I got the money." He told his son, "I never had a woman I didn't give something to." Jake even spoke to his granddaughter Erma about his out-of-town flings. He told her that he never had a regular mistress; the children that eventually found their way to him were the results of casual encounters. "When he was a cowboy he'd drive those cattle to Oklahoma City to be shipped, and he would stop along the way," Erma explains. "I think that's where he got all these babies, because these women would be at the station to meet him. And he told me, 'I never had a girlfriend—those women just got in the bed with me and I popped them.' He felt that it wasn't his fault because they shouldn't have made themselves so available."

Rose Simmons did feel it was her husband's fault—yet she forgave him. She forgave him the first time an unwed mother brought her a baby to raise from one of his illicit encounters. She forgave him the second time, too. And the third. And even the fourth. There's even a family joke on the subject: it is said that one of the babies which Rose unknowingly delivered as a midwife was her husband's son Jim.[12]

Somehow Rose Simmons made room in her home—and heart—for all four of her husband's "outside" kids. Even Eugene, whom Jake begot with Rose's younger sister Elizabeth, was welcome in the Simmons home.[13] "He never denied any of his children," recalls John Simmons. "And my mother didn't resent one of them."

Jake was, after all, an attentive husband: Rose's ten children attest to that. And despite his infidelity the couple was well matched. Their union was like the joining of two worlds. Jake was carefree, loose-natured, unsophisticated, talkative, generous, and trusting: the free-wheeling western cowboy. Rose was his family's domestic anchor. Well educated, unadventurous, strict, suspicious, and hard-edged, she was the quintessential frontier woman, running the couple's home with a relentless thrift that bordered on miserliness.

Over the years Rose and Jake used their home to raise four sons, six daughters, four "outside kids," a couple of destitute adopted boys, and a variety of other relatives, especially grandchildren. Jake's willingness to support some two dozen dependents probably stemmed from his own parentless adolescence and the example of Bluford Miller's benevolence. Yet he and Rose had little interest in making things easy for their children. Ranch life was steeped in hard work.

Every member of the family was part of its labor force, and there was no such thing as weekends off.

Until Jake and Rose's children were married they lived together in the family's original log home. The rambling house grew with the family. There were ten comfortably furnished rooms, including a few sitting rooms, an enormous dining room/kitchen, a well-stocked storage cellar, a bedroom for the parents, a bedroom for Jake Jr. and three of his brothers, two bedrooms for the daughters, and a "bunker" for the eight or so unmarried ranch hands who worked with the cattle. Rose was careful to keep her daughters and granddaughters out of the hands of the hired hands, or "boys." "The boys' room had a great big fireplace, and we wanted a lot to go in there where those boys were," recalls Ophelia Brown, who grew up in the house at the same time as Jake Jr. "We would have to get permission. The boys wasn't allowed to come to our room unless they came to make a fire, and get out."

The family spent much of their free time relaxing on the winding porch which surrounded their house. The area of porch connected to Jake and Rose's bedroom was fenced in. Jake enjoyed sitting there by himself slowly reading the newspaper when he found a few moments of leisure. Which was not often.

The household awoke before 4:00 A.M. Jake Sr. was the first up. If the children did not rise by themselves their mother woke them. During the winter if they kept sleeping Jake threw a bucket of water on the room's heating fire and left it to the freezing morning air to get them moving. More than a dozen Jersey cows were milked by the Simmons boys, and the horses, hogs, and chicken were fed. While they were out the girls prepared large country breakfasts of eggs, hash browns, rice, bacon, sausage, and hot biscuits with sorghum molasses, gravy, and butter.

After breakfast the morning's chores continued until it was time for school, then resumed in the afternoon. The boys helped the older hands with the cattle and property upkeep. Their father, an expert rider, taught them horsemanship skills early in their lives. Every morning the boys had to "ride the line," which meant patrolling the ranch's five hundred acres of grazing land for a break in the wire fence. John and Jake Jr. spent much of their youth in the saddle.[14]

The sale of cattle provided most of the family's income. It was hard, dangerous work. Cowboys had to transport, rope, doctor, breed, birth, and castrate the livestock. At the end of each fattening season Simmons and his men drove a few hundred head about four miles to a livestock lot in Haskell. The cattle were sold in advance to companies in St. Louis, Kansas City, or Oklahoma City. The rancher's herd usually filled three or four boxcars. Simmons traveled with the train to collect his payment personally, sometimes bringing one of his sons with him as a special treat. As his reputation grew he began brokering cattle for other ranchers and farmers. At one time he earned more money from other people's livestock than his own. In 1915 he was profiled in *Southern Workman,* a monthly publication of Hampton Institute (Booker T. Washington's alma mater). He explained that in the preceding year he grossed $5,000 selling his own cattle and netted more than $7,000 buying and selling for others.[15]

Even as a young child it was obvious that Jake Simmons, Jr., was destined for a life beyond his father's ranch. He enjoyed working with his mind, and developed an early aversion to agricultural labor. Studies and good grades provided a convenient escape from his more tedious chores. "Jake dodged and ducked all the hard work there was," recalls his brother John. "That makes a man study to keep from doing something [else]. As long as he was riding it was all right. When it came to walking he didn't want that."

When Jake Jr. wanted to get out of a chore he would begin it, then find an excuse to do something else. He was never one for lying. He constantly let out the tension of what was on his mind—often without regard to whether his listener wanted to hear it or not. At an unusually early age, Jake Jr. was able to communicate serious matters clearly, including what he wanted to do with his life. By the time he was five he began calling himself a man. "A man doesn't want to have to milk the cows at four-thirty every morning," he'd tell his brothers. "A man doesn't like being told what to do all the time."[16]

He was more careful with what he told his father. "Sassying Dad," as it was known, was as inconceivable as directly ignoring his orders. The child's "ducking and dodging" did not always succeed. When it failed, Jake shared the same fate as his six brothers: the fearsome strap. The strap, designed to sharpen razors, more frequently found its place against the backsides of misbehaving boys. It hung ominously on the doorknob of their parents' bedroom. Jake Sr. appreciated the swift effectiveness of his tribe's whipping-post justice. He never drew blood, but rarely quit before tears flowed. There were rules in the Simmons home, and when the boys disobeyed them (doing things like staying out all night) they knew what they could expect at home. John remembers trying to buffer the blows with an extra pair of pants, but his father always checked first to be sure his message was getting through.[17]

Rose also used the strap on her sons, but her whippings were more lenient than her husband's. To the girls, however, Rose was the more physically threatening of the two. "Grandma never got a strap," recalls Erma Threat. "But she'd back into you and WHACK—knock you in the mouth. Her pet thing was 'I'll make you swallow your teeth.' "Jake never raised his hand to the girls. Instead, he'd "call them to the carpet," a euphemism for his judgelike lectures on the porch. "He used down language," Erma explains, "not curses, but things like 'you're never gonna do nothin', you're just a slut, you're just a nobody.' He talked you down so low you'd rather him whip you than that tongue-lashing."

The Simmons ranch provided a livelihood for scores of people. In addition to family and hands, nine tenant farmer (sharecropper) families lived along the outlying area of his property. Jake Sr. was their local chief. During good times he profited by their labor, earning a 50-percent share on their crops of corn, oats, soybeans, peanuts, and sweet potatoes. When hard times hit he fed them and supplied work if they were unemployed. They relied on him for business advice, and he also helped provide their families with a local school and church. He was among the last of the county's black rural leaders. During the early 1950s, when Simmons's health finally forced him to leave the ranch and move to Muskogee, dozens of blacks in the Haskell region who had found security as part of his provincial estate also had to abandon the area for nearby cities.[18]

The poverty of Simmons's youth brought him a deep empathy for the needy. He rarely turned anyone away who needed a job. His ranch hands tended the cattle and horses, maintained the land, cut hay, planted and harvested crops, and handled construction and repairs. They were paid only about fifty cents a day, but the free room and board was better than most had ever known. Eight or more young men worked for Simmons full time. They stayed in the bunk room in back of the house or in a few small shacks he provided nearby.

The rancher was also accommodating when it came to loans, even when he knew they would never be repaid. "Left on her own, my grandmother wouldn't help anybody but who she liked," Erma Threat recalls. "He [Jake] didn't have to like you. He just knew that it was his duty, as a Christian, to supply the needs of the people who didn't have as much as he. There was no welfare, and many blacks were selling their farms and moving to the city, leaving a lot of boys between the devil and a hard place. If they came by he always said, 'I have this job for you'—like feeding the hogs or milking the cows or helping on the barn. Then he'd let them stay."

Even "day workers" were fed lunch and supper, eating in shifts at the family's eighteen-seat table. They added to the ranks of the ranch's twenty regular diners, keeping the girls constantly busy at the industrial-style six-foot-long, eight-burner stove. "It wasn't like you had to be a member of the family to eat," recalls Erma Threat. "If you walked in and we were eating, you ate. My grandfather wanted everyone to know they were always invited. It was kind of like an open house. He often fed hungry people and told us, 'We have and we have to share.' "

In addition to helping prepare meals the women of the house helped Rose can endless quantities of fruits and vegetables. Rose was obsessed with hoarding food. The family's basement resembled a Mormon survival cellar, with row upon row of thousands of home-grown and home "canned" (actually bottled with a pressure cooker) fruits, vegetables, and even chicken parts. Pork and beef were slaughtered en masse (Jake would butcher more than a dozen hogs at a time), smoked in the Simmons smokehouse, and also stored. Every jar was marked with its content and date of canning. The basement held enough to last through years of long winters. Rose never threw anything away. When she died, in 1947 (at ninety-two), relatives found jars dating back forty-five years.[19]

The farm's subsistence garden supplied the Simmonses with canned corn, peas, greens, sweet potatoes, cabbage, beets, okra, watermelon rind, and more. Wheat was taken to the mill and ground for flour. Clabbered milk was used to make yeast. Cucumbers and rendered into twenty-gallon barrels of different types of pickles, kraut, and relishes (including "cha-cha," a cabbage garnish still popular among Oklahoma blacks). From local trees came a plentiful supply of nuts, peaches, pears, apples, and berries. For sweetener, there was sorghum molasses and sugar cane. Milk filled enormous cans; cream and excess milk were sold to a buyer in Haskell, who shipped it off in a refrigerated rail car. Unsold cream was churned into butter or served with coffee.

Rose used an incubator to hatch chicks. The family kept more than a thousand chickens and hens—so many it was nearly impossible to walk out of the house without stepping over one. Turkeys and domestic geese were housed in tents, and fresh perch and catfish thrived in local ponds. Rose's food management was so efficient that despite all the people fed on the ranch each day the only items that needed to be bought from a store were tea, coffee, rice, salt, baking powder, and baking soda.

Bookkeeping at the ranch was a complicated matter. Because Rose was adept at "figurin,'" the family's business affairs were left to her. She managed money well and was fanatically thrifty. She even made lye out of ashes, then used the lye to make her own soap ("RS" was stamped on each bar). Like her parents and grandparents, she buried her valuables. Rose owned her own small herd of cattle, kept separate from her husband's. She hoarded the profits, keeping thousands of dollars in gold coins, as well as her fine silver and linens, tucked away in a secret cedar chest beneath her bedroom floor boards. The chest had a large padlock on it, and nobody—including her husband—knew where she kept the key. The couple's bedroom was strictly off-limits; some relatives lived in the home for years without seeing it.[20]

Rose also filled the shoes of a country doctor. As a midwife and herbalist she tended to the entire area without charge. A supply of clean rags was kept ready for child-delivery calls. She did not need a pharmacy to supply her medicine chest. Like many black Creeks she never shed the ancient knowledge of her tribe; from the Muskogee language she used with her husband (when they didn't want the children to understand) to the customary dishes she expertly prepared, Rose was as much a traditionalist as any of her Native American neighbors. She always knew where to find the natural compounds used by Creeks to treat virtually every disorder. Peach leaves were pounded up, boiled, and applied to the head to reduce fever. Slippery-elm bark was used as a poultice to reduce inflammation. Ginseng and ball willow cured pneumonia (known as "winter fever"), and life everlasting, an herb, was the favorite remedy for the common cold. Soot from the fireplace was applied to wounds to retard bleeding. To alleviate depression caused by a woman's period, a powerful stimulant, snakeroot, was made into tea and administered with a heavy dose of sugar and rum. Every spring Rose concocted another tea, from sassafras, for all the children in the household to drink, saying it would get "all the impurities out" of them. She also made a medicine out of rock candy and corn whiskey which, despite attempts to hide it, was regularly plundered by household raiders.[21]

Rose was not one for entertaining. She was introverted and quiet among those she did not know well (relatives considered this characteristic of a traditional Creek woman). She had few friends and rarely left the ranch. When visitors came she would change the immaculately clean "work" apron she always wore for a more formal "dress apron"—but she would not be seen in her own home without one. For a few minutes she would do her best to make small talk, then excuse herself for something that needed doing. Her far more congenial husband was always glad to carry on the conversation.[22]

Rose was most comfortable talking to young people She called it "training," and her conversation usually regarded the work at hand. If the girls tried to question the necessity of her insatiable demand for domestic chores she would warn, "Everything is warm now, but one day it's going to get cool and you and your husband are not going to get enough, and you will be trying to keep warm."[23]

Rose saw to it that everyone in her house worked. Youth was no excuse. Those too little to prepare food for canning washed jars—their small hands cleaned where larger hands could not fit. She was a harsh mistress when it came to having chores done properly. The bleached wood floors were scrubbed, not mopped. Working on their knees with a brush and lye soap, Rose's daughters and granddaughters kept the floors clean enough to eat from. When everything was spotless she would have another job lined up for them.

Rose did not think leisure time was good for young women. "An idle mind is the devil's workshop," she'd say. Her favorite bit of busy work was rendering scraps of excess cloth into quilts. Erma Threat sometimes resented the constant regimen. "I didn't want to sew, but I had to sew," she recalls. "Everybody had their own basket and had to piece quilts—you couldn't live in that house without you piecing quilts. You could not sit out and not do anything. You had to have something in your hand—you had to be piecing a quilt, embroidering, peeling peaches or doing something. When you got through you washed your hands and went to bed. And everything was ironed, even down to the washcloths and dish towels. I didn't understand why we had to go through such drudgery."

Once, after finishing a dress, Erma attempted a silent rebellion. She took her leftover scraps, needles, and pins, and swept them outside in the rain. Rose, ever-vigilant, spotted her. "Go get every one of those scraps!" she commanded. They were muddy and wet. Erma had to wash and iron them before piecing together yet another quilt.

Often the children tried to stay out of Rose's way. "People were more afraid of her than my grandfather," explains Johnnie Mae Simmons, another granddaughter. "He never spoke out too loud. My grandmother [also] had a voice that wasn't so loud, but it was so distinct and it seemed to have more authority to it. She was more or less the boss—everybody would do what she wanted done."

Erma Threat believes that Rose was the "foundation" for her husband. "She would keep him abreast of everything," Erma recalls. "They would go into this little room, and she would tell him everything that went on that he didn't see—she'd say those boys [the ranch hands] didn't water the horses, they didn't curry them, they didn't do so and so. Grandpa was the kind of man that didn't look at details; but she did. Sometimes he didn't respond to it right then. He would let it add up, until he got real mad. Then people would wonder how he got all this information."

Rose made sure her children received a different kind of education. Both boys and girls were taught to shoot guns. From the age of ten they began firing at tin cans set on fences. If they got good enough, they learned to shoot the cans in midair. Every member of the family was able to defend the ranch.[24]

Instructing came naturally to Rose. She taught Johnnie Mae Simmons to write before the granddaughter ever went to school. Johnnie Mae would help gather eggs by the chicken coop, then they would sit together outside while Rose showed her how to spell her name with a stick in the dirt. Rose and Jake both felt strongly about the need for education—for girls as well as boys. Because there were dozens of children living around the ranch community the Simmonses decided to build their own school. After winning the support of the superintendent of Indian Territory, whose administration supplied the teacher, they and two neighbors constructed a $500 two-room schoolhouse.

The community also used the small school as its Baptist church until Jake built his own. He did not take religion very seriously until after he turned fifty; a tent revival meeting inspired him to testify and become baptized. In 1919 he borrowed $1,400 and erected a large wooden church on his land, complete with chapel, meeting room, and picnic grounds. It was dubbed "Simmons Baptist Chapel" and is still in use today.[25]

Church services were the only thing that got Rose to wear a hat. To Jake Sr., dressing up on Sunday morning meant putting on a better-pressed variation of the same outfit he always wore: a khaki-shirt-and-pants outfit, with the top button buttoned and taut suspenders. (John Simmons recalls that only once did he see the rancher in a suit. At the time Rose called the boy in to help fix his father's tie. John scarcely recognized him.)

Apart from church, baseball, and horseshoe pitching, breaks in the monotony of farm work came from special events, and biannual junkets into the city of Muskogee. Jake would hitch up his buggy, pile his excited children in the back, and ride them into the town of Haskell, where they would catch a local train to Muskogee, then the largest city in the Territory. Although a direct drive in an early-model automobile would have gotten him there quicker, he delayed his first purchase of a car for many years. He was proud of his fine "uptown" horse buggy, and took great pleasure driving it to Haskell each morning to pick up the mail. He bought a new-model buggy every few years, and kept himself up to date with rubber tires and the latest fold-down tops. The buggy was pulled by his best horses, adorned with beautiful harnesses. Jake was a horseman to the very end. Although he lived until 1955, he never learned to drive. He eventually bought a car, but it was for his children and grandchildren to use, and occasionally chauffeur for him.[26]

The drive from the Simmons ranch to Muskogee can now be made in half an hour, but the distance by buggy and train eighty years ago sometimes made it necessary to stay in the city overnight. The family would check in at the all-black Rebecca Turner Hotel, then hit the town. Muskogee grew quickly after the turn of the century as non-Creeks were allowed to settle in the area. The city's population increased from just a few hundred in the 1870s to 4,254 in 1900 and 14,814 in 1907. The black district of Muskogee also grew. By 1907 it contained the most extensive network of black-owned businesses in the Southwest. Muskogee's version of Harlem's 125th Street was Second Street, a wide avenue in the heart of downtown. It had the city's convention hall, an amusement center, and an assortment of black and integrated businesses, including a black-owned bank, jewelry store, commodities brokerage house, and doctors, lawyers, dentists, restaurants, boardinghouses, and barber shops.[27]

The Simmonses' first stop on Second Street would be the highly successful black-owned stores. Jake dragged his sons into Adam's, a men's clothing store, and outfitted them with new shoes, and whatever clothes Rose and her helpers were unable to make themselves. Rose and the girls went to Adam's sister store, T. J. Elliot's, the finest woman's clothing store in town. Rose rarely bought dresses, or anything she could make herself. But she did buy the best soft leather shoes available—always black, high-topped, and thoroughly conservative.

For Jake Jr. and his brothers the fun began when the shopping ended and they were on their own. Jake Sr. often timed their trips to Muskogee to coincide with county or state fairs. Even now such fairs remain the most popular form of mass entertainment in Oklahoma. There were animal and agricultural exhibits, horse races, carnival rides, and competitions. Rose regularly took home blue ribbons as champion of her local "'farm federated club," winning awards for quilts, canned goods, hens, hogs, and cows.[28]

The boys didn't stick around for the exhibits. The first chance they got they joined up with a pack of other kids, sometimes disappearing with their girlfriends. Although they felt slightly awed by the city they never showed it. "Dad would cut us loose," John recalls. "We were loose to do whatever we want exceptin' kill anyone."

* * *

By 1890 there was no place in America where blacks could lead as dignified a lifestyle as they did in the Creek nation. The two races most often oppressed by the white American majority and its government were left alone together, like a pair of outcasts lucky to find themselves exiled beyond the reach of a tyrannical ruler. Black votes often swayed tribal elections, and Creek politicians made sure their rights were respected. Black equality was a result of the tribe's success in keeping whites out. Although lax immigration laws allowed Caucasians to form overwhelming majorities in the neighboring Choctaw and Chickasaw nations, Creek tribesmen, one-third of whom were black, outnumbered whites five to one. But the tribe's power to restrict illegal settlers steadily eroded. Although the U.S. opened up nearby lands to white homesteaders, it was not enough to satisfy illegal settlers, or "boomers," who demanded an end to what they called the territory's "domination" by tribal government. The federal government's 1832 treaty with the Creeks had assured them that the western nation was theirs to govern as they liked, "as long as the waters run." Yet by the end of the nineteenth century U.S. officials openly encouraged illegal white settlement. When this led to interracial land disputes the federal government came up with a simple solution: dissolve the Creek government and end the tribe's communal system of land ownership. Each tribesman, according to the plan, was to be allotted 160 acres, which he would be free to sell to whites. The hundreds of thousands of Caucasians already living in the surrounding area would then be able to form a state and become American citizens.[29]

Members of the Creek tribe voted twice during the 1890s on the allotment proposal. Both times it was overwhelmingly defeated, with blacks voting alongside red tribesmen to maintain their Indian nation.[30] The tribe's chief appealed to "the great American people" to "keep their treaties, respect our weakness, render encouragement, and strengthen our hand in this supreme effort we are making for the survival of our race." Yet few governments respect weakness, and America's was no exception. They ignored the Creeks' will and set up a commission to parcel out the last vestiges of a once-great Indian nation.[31]

An unexpected side effect of the government's insistence on individual land allotments was the largest transfer of land to African-Americans in history. Some 37 percent of the Creek tribe consisted of blacks, and their allotments eventually amounted to more than one million acres. Allotments for blacks living among other Indian Territory tribes totaled another million acres.[32]

Such a large postslavery land distribution was unique in North America. After the Civil War, abolitionists had been completely unsuccessful in convincing Congress to provide freedmen with land confiscated from the Confederacy. It was too much like giving capital to workers, and after a short time the government even forced freedmen occupying abandoned land to relinquish it to the original owners. A relatively tiny amount of land did reach black homesteaders, but all in all the four million slaves freed throughout the U.S. probably received less land than the total received by 23,400 blacks in Indian Territory.[33]

The difference this made to Indian blacks like the Simmonses is no small matter. In his exhaustive 1935 book *Black Reconstruction,* W. E. B. Du Bois had this to say about the absence of land distribution in the United States:

> To emancipate four million laborers whose labor had been owned, and separate them from the land upon which they had worked for nearly two and a half centuries, was an operation such as no modern country had for a moment attempted or contemplated. The German and English and French serf, the Italian and Russian serf, were, on emancipation, given definite rights in the land. Only the American Negro slave was emancipated without such rights and in the end this spelled for him the continuation of slavery.
>
> . . . Surprise and ridicule has often been voiced concerning this demand of Negroes for land . . . as far as the Negroes were concerned . . . anything less than this was an economic farce. On the other hand, to have given each of the million Negro free families a forty-acre freehold would have made a basis of real democracy in the United States that might easily have transformed the modern world.[34]

Blacks in the Creek nation were born lucky: they received not just forty acres, but a 160-acre "freehold." This may not have transformed the entire world, but it certainly altered the world of Jake Jr.'s youth. Like 6,808 other Creek blacks born before the March 4, 1907, cutoff date, Simmons was entitled to a birthright of 160 acres of tribal land. So were his brothers, sisters, and parents; the family's allotments totaled well over a thousand acres.[35]

In an agrarian community the possession of land meant the ownership of the means of production. In Indian Territory a class of black landowners was far better able than their wage-earning countrymen to control its destiny. Once the tribal government was killed, men like Jake Simmons, Sr., began to appreciate individual ownership of land. He knew that if anything was universally protected for all people in the United States it was private property. The thousands of white émigrés arriving in the Territory each year agonized over the privileges and "uppityness" of the Indian blacks. But they did it outside Simmons's fence, and well beyond his family's earshot.

For many of the black tribesmen the fences did not stay up for long. Although they had previously been allowed to use all the tribal land on which they lived, ownership brought a new responsibility to thousands of farmers who had never had more than a few dollars to call their own. There was an onslaught of "grafters," con men who, after softening the new landowners with drink, were often able to buy up an entire family's allotment just by offering an unfamiliar quantity of flashy cash. When that didn't work they tried more devious means. A grafter would give an ignorant title holder an instant mortgage for a trendy new horse and buggy, then return in a short time, demanding a first payment. When the freedman did not have it his farm was bought out from under him. Another popular scam was to hire young black men from outside the area to pursue and marry single women of legal age. Upon assuming possession of their wives' acreage the "husbands" would immediately sell it for a nominal fee to their grafter bosses, then disappear.

The white courts turned a blind eye to the victims of such schemes. Most Caucasian settlers, hailing from states where blacks assumed a far more servile position, felt the grafters were performing a service to the community. They despised the landowning freedmen, especially the Creeks. White supremacists wanted the darker race to be destitute and helpless, fawning for mercy and starvation-wage employment. Such men "knew their place." But in Indian Territory they felt threatened by a class of blacks who defied stereotypes. Local newspaper stirred such bigotry with headlines warning of nefarious "race wars" and exposés about "insolent" Creek Negroes with money. In 1908 the *Oklahoma City Times* ran a prominent article headlined "The Negro Creek: A Study in Criminology." The newspaper assailed "the very thin veneer of civilization which overlies too many of the Negro-Creek citizens" and noted that because "the barbarity of both races showed up in its true colors . . . the Negro Creek has all the natural-born callousness of the genuine criminal . . . [and is] the most dangerous man on the American continent today."[36]

In a sense Creek freedmen like Jake Simmons, Sr., did pose a threat to the racial-supremacist system of the white majority. They were able to provide their children with a sense of dignity antithetical to the submissive role to which less independent blacks often succumbed. One of Jake and Rose Simmons's greatest achievements was that they never sold a land allotment—no small feat at the time. According to historian Angie Debo, by 1906 grafters "had secured practically all the freedman land from which restrictions were removed in 1904."[37]

That the Simmonses were able to resist the connivances of the grafters might well be attributed to Rose. "In some ways she was his backbone," explains Erma Threat. "He was very generous and communal. She was very miserly. He was so naïve he once signed guardianship for a friend's child, and when the boy got older he sued him."

Because Rose was far more literate than Jake and responsible for the ranch's bookkeeping, any papers her husband signed were first brought before her suspicious eyes. A few hundred dollars or a sporty buggy would do little to tempt a woman raised in relative affluence, who regularly lectured her children, "Never want what other people have 'cause you never know how they got it."[38] Nor would it skewer the judgment of a man like Jake Sr., who was accustomed to handling thousands of dollars for cattle transactions.

For successful ranchers like Simmons, when the flash of cash proved ineffectual there was a much more gradual method of parting the black Creek from his land: drowning him in debt. A local white banker made a fortune by first extending credit to black ranchers to increase and upgrade their herds, then foreclosing on the land when the vagaries of the cattle industry turned against them. Simmons also borrowed from this banker, the most active in the area. He struggled with debt much of his life, but always stayed a few steps ahead of the next payment—the difference between a prosperous cattleman and a bankrupt sharecropper.[39]

The Simmonses even managed to add to their land holdings. Most family estates were broken up when children reached legal age and sold their allotments to raise cash or pay off debts. Simmons retained his estate by buying land from whichever children insisted on selling it, even when he had to borrow money to do so. Although Jake and Rose registered early and got a number of their family's land allotments together, some of the younger children were assigned property far from the family ranch. Jake became adept at swapping pieces and exercising options on bordering property, always with an eye toward consolidating his "block." In 1915 a land abstract showed members of the family holding eight contiguous 160-acre sections.[40] Simmons also bought out some of his neighbors who had quit farming, so that by this time his holdings approached 1,800 acres. When he died in 1955 he left a will which forbade any beneficiary to sell the land, unless it was to a close relative.[41] The family of his son, Jake Jr., maintains control of most of this property today.

2

"I WANT TO BE AN OILMAN"

In the long run, the world is going to have the best, and any difference in race, religion, or previous history will not long keep the world from what it wants.[1]
—Booker T. Washington, Up from Slavery

The unique advantages enjoyed by black Creeks like Jake Simmons did not go unnoticed by African-Americans in other parts of the United States. Oklahoma nearly became the first black-ruled state in the nation.

In the late 1870s, with the final defeat of a pro–civil rights cartel in Congress, thousands of blacks in southern states packed their bags and headed west. After a decade of Reconstruction which saw twenty blacks elected to Congress and hundreds more serve as state legislators, the nation's politicians had turned their backs on freedmen. As occupying Union armies marched out of the Confederate states, many African-Americans hurried to escape the inevitable reign of terror which followed.

A great migration of people determined to find a life beyond the cotton fields their parents had slaved in turned their eyes toward Kansas and Oklahoma. In 1879 alone nearly 60,000 of these "Exodusters" moved to Kansas. Many more would have joined them, but southern planters blockaded the Mississippi River and stopped African-Americans from leaving. (To avoid depleting their cheap labor supply, planters used sharecropper indebtedness as an excuse to detain forcefully the would-be émigrés.)[2]

Kansas was far more progressive than any of the southern states, but land was a scarce commodity for the cash-poor Exodusters. Many African-American leaders saw the vacant government-owned "unassigned" lands of Indian Territory (west of the sovereign Indian nations of the Five Civilized Tribes) as the ultimate resettlement destination for members of their race.

These activists lobbied top federal officials in Washington to open the unassigned lands to former slaves from across the United States. The government instead decided to distribute the land to homesteaders of all races. The black "colonizers" were undeterred. The possibilities of a new district with no existing white power structure ignited their imagination. They believed that by convincing enough of the five million oppressed blacks in the South to move to Oklahoma, they could become the racial majority of the voting public, elect their own politicians, and create a democratic system of social and economic justice.

The colonizers were led by the former auditor of the state of Kansas, a distinguished-looking African-American named Edwin McCabe. His agents swept the country, appearing in black churches and public halls across the South, sending letters to newspapers, preachers, and schoolteachers. McCabe promised that the capital of the black state would be Langston, a prosperous, fast-growing town which would become "the Negro's refuge from lynching, burning at the stake, and other lawlessness."[3] Exaggerated rumors of McCabe's activities stirred the worst fears of white settlers. On March 1, 1890, an article in the *New York Times* noted, "If the black population could be distributed evenly over the United States it would not constitute a social or a political danger. But an exclusively or overwhelming Negro settlement in any part of the country is, to all intents and purposes, a camp of savages."[4]

Other newspapers predicted imminent race war and reported frequent threats on McCabe's life. But warnings about hundreds of thousands of black colonizers never materialized. Emigrants were risk-takers, and risk-takers are always in the minority. By 1900 the region's black settlers were outnumbered by whites more than ten to one, and the dream of a black state died.[5]

The consequences which the colonization leaders feared would accompany a white majority state were fast in coming. As the merger of the individual Indian nations and Oklahoma Territory loomed closer, blacks shuddered at the prospect of a white-dominated state government. Even before statehood all-white towns like Norman and Sapulpa began banning Negroes, sometimes purging entire counties of blacks who had settled there first. Only in Muskogee County, where Simmons was raised and African-Americans constituted nearly one-fourth of the population, did black dignity make a stand.

For black Indians Muskogee was a last stronghold in an ever-rising sea of racial hatred. As late as 1905, three blacks were elected to the City Council. That year *Muskogee Cimeter* editor William H. Twine defined their struggle in an editorial: "The liberty-loving people regardless of race or color will not stand for the kind of oppression here that reconstructed rebels desire. . . . Some of us have made our last move and we propose to stand on this ground where we have our homes and our investments until hell freezes over and then fight the devils on ice. . . . The Indian Territory is the last stand the Negro of America can make as a pioneer and we propose to let it go down in history that the stand was made here."[6]

A few years later Oklahoma became a state. The Democratic Party, arguing that blacks would overrun the state if a Republican government took power, won a resounding victory in Oklahoma's first election. The very first law enacted by the state legislature was a segregated railways bill. More than five hundred railway depots had to be altered to make separate waiting rooms. To leave no doubt who fell into what caste, racial distinctions were spelled out by law, and kept simple: everyone was included as a member of the white race, except those with "African" blood.[7]

Oklahoma's lawmakers felt a "white man's burden" to undo the racial mixing prevalent in the "lawless" territories. (By one estimate all but a handful of Creek families contained some amount of Negro blood.)[8] The state passed a strong antimiscegenation bill, making marriage between races a felony punishable by as many as five years in prison (ministers performing such ceremonies were also subject to felony charges). Since Oklahoma's government regarded Native Americans as whites the legislature used the law to push Indians into the white man's path of racism.

Their effort was largely successful. John W. Simmons was eight years old when Oklahoma became a state. "I wished it'd have stayed a territory," he says. "Governor Haskell and the legislature stopped the intermarriage [between blacks and Indians] with a law, and lots of whites married into Indian families. After the Indians went on the white side they cut loose on the black man completely. I had a lot of friends, Indian boys and girls. They got kind of honkie about it. I never let on as though I knew why, but you could read between the lines: association brings about assimilation."

In 1910 the Democrats took their boldest step against the state's blacks: they took away their right to vote through a grandfather-clause amendment to the Oklahoma constitution. The law restricted voting to those who were eligible to vote (anywhere) on January 1, 1866, and their descendants, as well as foreign immigrants. All others—meaning all blacks—were required to take a "literacy" test. Administered by local registrars determined to keep blacks from voting, in most counties these were often impossible to pass. By 1915 the state's racists had even managed to outdo their southern neighbors: Oklahoma achieved the dubious distinction of becoming the first state requiring separate telephone booths for blacks and whites.[9]

Thousands of black Oklahomans tried to escape the humiliation of segregation by keeping white people out of their lives. For those without a ranch from which to fence off an ugly world, there were the black towns. Oklahoma contained well over two dozen all-black towns, more than any other state in America. The larger black towns were entirely self-sufficient. African-Americans ran their own baseball teams, Masonic lodges, newspapers, banks, businesses, local governments, schools, and law-enforcement agencies.[10]

Outside the black towns African-Americans practiced a strange form of segregation within their own community. The unique heritage of the Indian freedmen, especially the Creeks, was a source of pride to the old time prestatehood citizens. Some looked with disdain upon the newly arrived African-American immigrants, who continued moving to the black towns and Oklahoma cities in large numbers during the early part of the twentieth century. While researching his 1940 graduate thesis, University of Oklahoma historian Sigmund Sameth interviewed a dozen elderly Creek freedmen who resented the "state" blacks, or "Watchina," as they were known in the Creek language, for the caste system that evolved after statehood. "What do I need to mix with them state folk for?" one told him. "I was eating out of the same pot with the Indians, going anywhere in this country I wanted to, while they was still licking their master's boots in Texas."[11]

White Oklahomans were fully aware of the difference between state and Creek blacks. In 1940 Sameth surveyed three large white farms in eastern Oklahoma which used black laborers and found that all three refused to hire Creek freedman. When he asked a Creek black why this was, the man replied, "We're too high-natured. They can't kick us around like Alabama or Mississippi niggers."[12]

Many such "natives" kept their children away from the children of state blacks, and discouraged their intermarriage. Jake Simmons, Sr., was not one of them. Like skin color, geographic background had little to do with how he regarded a child's spouse. Simmons was more concerned with character, and whether or not they would work hard on a farm.[13]

The farm meant everything to Simmons. He spent little energy fighting the system. Instead he withdrew from it, into the one world which guaranteed independence for him and his family. The rancher believed that only through the ownership of land could an African-American maintain his dignity in a racist society. To young and old alike he offered the same advice. "Get you a home," he would say. "Buy you some land and get you a home."[14]

Many of Oklahoma's blacks looked for that home outside the new state. Disgust over the local government's racism convinced a number of idealistic settlers that greener pastures would only be found in other countries. Hundreds of them, including Edwin McCabe, moved to Canada.[15] Others attempted a more radical alternative—moving back to Africa.

The first back-to-Africa movement to reach Oklahoma was started in 1913 by a dynamic orator from the Gold Coast (now Ghana) known as Chief Sam. He went bankrupt without bringing most of his investors to the African paradise he promised them.[16] A few years later the message of Marcus Garvey, the "Black Moses," found a multitude of eager followers among disenchanted Oklahomans—as well as more than a million other blacks around the world. Garvey's Universal Negro Improvement Association stressed economic autonomy, political empowerment, cultural affirmation, and black liberation theology. His organization, although poorly managed, ran a steamship line, stores, small factories (which made very popular Negro dolls), and a travel bureau. A number of Simmons's friends in nearby Haskell were committed Garveyites whose quest for economic autonomy led them to build the town's first cotton gin, then sell their possessions and emigrate to Liberia.[17] The idea never appealed to Simmons. His thousand acres brought him all the autonomy he needed.

Land was also important to Jake Jr., but in a different way. To him, land represented an asset, not a lifestyle. By the time he was ten years old young Jake had made up his mind never to be a farmer. Sometime in 1911 his father assembled the boys in front of an open potbellied stove to talk to them about their futures. He asked each what they wanted to do with their lives. Johnny and the others answered like most young boys would: a cowboy, a marshal, a farmer. When the question came to Jake Jr., his response was as confident and unequivocal as it was shocking. "I want to be an oilman," he said.[18]

During the mid-nineteenth century, Native Americans from the Five Tribes discovered open oil springs in Indian Territory. They had no use for the substance as anything but a medicine, a remedy for chronic diseases like rheumatism and dropsy. The idea of cashing in on it never occurred to them. Until the turn of the century the fledgling oil industry was limited to the eastern states of Pennsylvania and Ohio, where it was ruthlessly dominated by John D. Rockefeller. In 1901 Pittsburgh tycoon William Mellon changed all that by striking the biggest gusher ever recorded at the "Spindletop" well in Texas. Hundreds of oil speculators descended upon Texas and soon spilled over into Oklahoma. In 1905 enormous reserves were discovered at Glenn Pool, near the small town of Tulsa, in the Creek nation. The oil boom had begun. By 1907 it was the new state of Oklahoma—not Texas—which led the nation in oil production.[19]

Overnight, Indian allotments which had been regarded as cheap grazing land became potential money farms. In 1914, at the age of twenty-one, John Paul Getty made his first fortune with a $500 lease on Creek land just outside of Muskogee. Around Glenn Pool, investors gobbled up every available inch of ground, offering as much as $43,000 (worth nearly ten times as much in current dollars) just for the right to drill on a Creek tribesman's allotment. The city of Tulsa grew from a population of 1,930 in 1900 to 18,132 in 1910 and 72,075 in 1920. For the lucky oil-rich Native Americans and Creek freedmen, many of whom had never seen a fifty-dollar gold piece, the one-eighth "royalty" interest which came with the lease payment meant monthly checks of thousands of dollars. Eastern Oklahoma was suddenly transformed from an agrarian hinterland to a kingdom of palatial mansions and overnight millionaires.[20]

To an impressionable young man like Jake Simmons, Jr., the oil boom meant the sight of local black and red Creeks wearing fine clothes, building lavish homes, and racing around in wondrous new horseless carriages. And unlike his father, who was constantly using his income to make mortgage payments on various parcels of land, everything the oil-rich had was theirs to spend. "Oklahoma was having a tremendous oil boom, and it was attracting everyone's attention," explains Jake Jr.'s son Donald, who currently runs the family oil royalty company in Muskogee. "At this time there was no income tax in the United States. When they found oil, they would run the wells wide open, and a person got three or four thousand dollars a month. When you see this kind of wealth the opportunity of the oil business really dawns on a person, especially if you're working on a farm with your dad, getting up at four o'clock in the morning and working till dark seven days a week."

Although no black had ever gotten anywhere in the oil industry Jake's father made no effort to dissuade him. The boy did not want to be told what to be, just as he disliked being told what to do. His lack of enthusiasm for farm work was apparent to Jake Sr., who assigned chores like milking the cows to his other children. Young Jake's excellent grades excused his tardiness at many menial tasks, and he was always able to talk a good game. He was never the slightest bit intimidated by adults, and used every opportunity to barrage them with questions, especially about business. Authority came with early maturity. "He always wanted to be a man," remembers John Simmons, who took life far less seriously than his brother. "He was a menacing thing. He believed in bossing all the time."

Jake Jr. was easiest to get along with when he was on a horse. Riding was one of the few things he truly enjoyed about living on the farm. Horses were to Simmons's sons what sports cars are to today's teenagers; Jake and John would sit up late at night talking about their horses and riding plans. Their father encouraged his boys to ride—learning to control a horse was good for the character. The family had a large corral, a cavernous barn, and two additional "outbuildings." They kept three to five dozen horses on the ranch, in pasture and stables. Most were workhorses for wagons and machinery, but a good number were for riding, racing, and rodeo.

In 1909 Jake Sr. started a local rodeo sideshow. A provincial version of Oklahoma's famous traveling "101 Ranch," it became known as "Jake Simmons's Cowboys." The group consisted of his sons, ranch hands, and neighbors. Nearly all were African-Americans. They performed annually at large fairs in the area. The boys carefully groomed their steeds and dressed them in equestrian finery, then showed off for enthusiastic crowds, riding wild "broncs" and performing lasso tricks, making the ropes dance in midair.

Jake Jr. was an excellent rider. From the age of seven he trained his father's racing horses. By the time he was nine he was a jockey, and regularly took first prize in races at the Muskogee fairgrounds.[21]

In many respects life was easier for Rose and Jake's sons than it was for their daughters. Like many parents of their day they allowed the boys far more freedom than the girls, upon whom they relied for constant domestic service. Also, Rose's old-fashioned thrift became embarrassing, especially to her younger children. Even after the Depression she continued dressing grandchildren in dresses and bloomers made from printed feed sacks. What other kids mocked as "mammy-made" was about all Rose allowed her girls to wear. She refused to spend money on anything they could make themselves with the help of a few basic patterns and an occasional purchase of good cloth and lace. She upgraded her sewing machine every few years to the latest nonelectric treadle model, and made sure her family got its money's worth. When Erma Threat was preparing to go to college in Kansas she asked her grandmother for a new coat. "You don't need two coats," Rose snapped. "You can only wear one at a time."

Jake Sr., on the other hand, was a pushover. When Erma came to him for a second opinion he sneaked her into Muskogee. "He took me to a shop and spent hundreds of dollars," Erma remembers cheerfully. "She almost had a heart attack."

The real trouble for the Simmons girls came when they reached young adulthood. On the one hand they lived in a traditional society where they were expected to become conventional brides. On the other hand they were unusually well educated and reared to fend for themselves. Given the world in which they lived it was a volatile combination.

It was the Simmonses' oldest daughter, Laura, who caused the biggest uproar. Laura was a good-natured woman with a hot temper. In 1911 she became a murderer.

Ophelia "Prock" Brown, Laura's daughter, was with her mother when the incident occurred. Because she retains the family's rural tradition more than any of the girls raised on the ranch, Ophelia is a good person to discuss early statehood days with. She lives in the same weather-stained house in which she was born eighty-three years ago, upon her mother's original Creek allotment (down the dirt road from where the Simmons ranch once lay). Ophelia is a guiding force at the nearby Simmons Baptist Chapel, attended every other Sunday by a dwindling but loyal group of older members (their minister alternates between two rural churches). I interviewed Ophelia in the fenced-in yard outside her house. She told me that many of her friends mistakenly think she is wealthy because "I can make do with little through Christ." With a thick rural Oklahoma accent she related this story of how she came to spend her adolescence with her grandparents while her mother served time in jail.

Sam Jackson, Ophelia's father, was unfaithful to his wife, Laura, and everyone knew it. Using his brother's house down the road as a base, he carried on an indiscreet affair with a woman named Rosie. When Ophelia was six years old and her sister was seven their mother grabbed a shotgun and decided to put a stop to it. "Mother put our coats on and drug us out in that snow," Ophelia remembers. "I said, 'Mother, where are we going?' She said, 'Daddy's up there with Rosie and I'm going up there and kill him.' Honey, it was cold, there was snow on the ground. When we got to the gate outside the house, we were just standing there. Mother said, 'Sam, you'd better come out there, you hear me? Sam, I know you're in there with Rosie.' Rosie came to the door—great big old tall stocky lady, she placed herself in the door and she said, 'Yes, he's in here—you come and get him.' My mother just decided, 'Well, big lady, maybe you want to take the consequences.' Mother had a single barrel. Mother caught that girl and Mother let her go right down in the dirt. She killed her."

Sam, meanwhile, had already jumped out the rear door and run back to his own house ten minutes away. "My mother just caught us by the hand and turned us round and brought us back home," Ophelia continued. "When we got home my dad was sitting up there at the house, he beat us back. Mother was so aggravated with him she unscrewed the burner off the oil lamp and threw it on him and was going to set him on fire. He ran out. I don't think he caught fire because we didn't burn the house down. My mother got in trouble and then my grandparents kept us."

Only one of the Simmons children served time for murder. The trouble he had with his other daughters was similar to that of many parents today: unwanted adolescent pregnancy.

The Simmonses were adamant that all fourteen of their children, both boys and girls, have a decent education—an unusual ambition for a rural Oklahoma family at this time. To a southwesterner at the turn of the century this meant attending an accredited secondary school, roughly equivalent to a combination high school and two-year college. Such schools were segregated, and few in number. To attend them students generally had to travel considerable distances and live in dormitories.

Most of the Simmons girls were sent to an all-girls teachers' college in Kansas. The school offered a six-year program beginning after the eighth grade. Jake and Rose sent three of their daughters there in the early 1900s, expecting the strict institution to assume the role of surrogate parent. After a few years, Jake received a notice that the two younger teenage girls, Dora and Oda, were pregnant. He rushed north to Kansas, pulled all three daughters out of college, and brought them home. In later years he sent other granddaughters to college with the stern warning, "If you mess up and come back you'll get the hoe—because that's the only skill you'll have."[22]

Even on the ranch Jake had trouble keeping an eye on his daughters. He forbade them to sleep anywhere but home; no exceptions were made. Girls were ordered not to engage in long courtships. After they had been seeing a boy for one month, Simmons demanded to know what their "program" was. If they liked the young man they were instructed to marry. If they did not marry him, they had to break off the relationship before it became too intimate.[23] Despite Jake Sr.'s own illustrious record he also laid down rules of conduct for his boys—at least on a local level. John Simmons recalls, "Dad had a philosophy that no girl would come up pregnant in that community by his children; if she did, you're a dead man, or you got to marry her."

When eighteen-year-old Oda ran off with a local boy after a picnic and stayed out all night, her father, in a fit of rage, grabbed a gun and rushed out after them, determined to render the young man into another notch on his pistol handle. "Let him alone," Rose yelled, dragging her husband back. "He's grown." The young elopers took a wagon to McAlester, more than seventy miles to the south, and were not heard from for a year. When they returned Simmons forgave and embraced them both.[24]

As is often the case it was the most obedient daughter who suffered most from her father's attempt to control his girls. Erma Threat's mother, Cora, was the Simmonses' second child. Although she was not pregnant she was yanked out of college in Kansas along with her sisters. This broke her heart because she had been in love with a senior at a nearby men's college who wanted to marry her. When Cora's college boyfriend heard that her father was coming to bring her home, he asked her to run away with him and get married. Unwilling to defy her father, Cora refused, a decision she regretted for the rest of her life.

Cora, who was nearing twenty at the time, wanted to keep seeing her college sweetheart (who eventually became a professor), but Jake would not hear of it. "He felt black teachers were paupers because they didn't have money and land," says Erma Threat. "My grandfather thought if you couldn't farm you really wasn't anybody."

Jake selected what he considered to be a better mate for his daughter. He'd recently allowed a young sharecropper from Texas named Will Brown to move onto his property. Brown was a hard worker, and Jake was impressed that he was able to support his parents on a half share. "You ought to marry him," Simmons instructed Cora when she returned from Kansas. "He's a good farmer and will make you a good living."

So Cora married Will Brown, and her father helped the couple set up a farm on Cora's 160-acre allotment. During their fifteen-year marriage they had five children, but Brown died suddenly of liver cancer at the age of forty-seven. Cora fell in love with another man of her choosing named Samuel Barnes, an educated cabinetmaker, cobbler, and musician. Barnes moved in and had two children with Cora, but the marriage was short-lived. Because his new son-in-law knew nothing about farming, Simmons treated him disrespectfully and tried to run his life. Barnes became fed up. He asked Cora to rent out her land and move with him to Muskogee, where he could play music and find work as a cobbler. Jake ordered his daughter to stay put. Again, she did what he said. Barnes moved to the city and got a job repairing shoes. The couple spent the rest of their lives apart, although Cora left the farm to take care of him whenever he was sick. She was unable to manage her land herself, so her father ran it for her. The Simmonses also brought Cora's younger children into their home, where they could "train" and rear them. Late in his life Jake Sr. admitted to granddaughter Erma that he regretted forcing Cora to leave school.[25]

Erma Threat has mixed feelings about life with Jake and Rose Simmons. "We had a neighbor named Richard Tucker who was down-to-earth and didn't want land and money as much as my grandfather," she says. "His boys didn't have to wake up at four and work so hard. My grandfather would say, 'He's not a good example of what black people should be aspiring for; he's not a successful man.' But to me they were always having the best time, always having picnics and fun things." Erma is still a bit resentful, still balancing the pressures of her youth with the prosperity of her adult life. A large, forceful woman in her mid-sixties who looks considerably younger, Threat retired from a successful teaching career and now runs a thriving Avon sales business out of her impressive home a few blocks from the state capitol in Oklahoma City. Erma's grandmother taught her the value of a dollar and she does not ignore the positive aspects of her upbringing. "My grandparents showed me more what life was really like. You've got to have a certain amount of discipline, a certain amount of routine, a certain amount of stick-to-itiveness. Now I appreciate that, because I believe that training made me what I am today."

* * *

Jake Simmons, Jr., also learned the value of hard work from his parents. He admired the results of their labor but did not want to follow in their path. He wanted to work with his mind, not his back, to use his hands for skilled, not menial labor. He moved quickly through seventh grade in the local primary school, surpassing his brothers and sisters, and finishing when he was about eleven years old. He and John, his closest brother in age, were then sent off to secondary school at the Tullahassee Mission School. Two of their older brothers had gone to school in Kansas and at Langston, but Jake and Johnny stayed in eastern Oklahoma, perhaps because of the trouble their pregnant sisters had gotten into. Tullahassee was more than a day's ride from the Simmons ranch, so the brothers lived in a dormitory, returning home only on special occasions and summers.

Tullahassee was a Baptist vocational school in which students worked to pay part of their board. As the oldest and best black school in the Creek nation, it accommodated more than one hundred students and was overcrowded. After their first year Johnny was sent to a different school in Muskogee. Jake, with stronger grades, was allowed to return. He stayed at Tullahassee another year, and would have continued there had Booker T. Washington not visited Oklahoma and changed the course of his life.[26]

When Booker T. Washington visited Muskogee in the summer of 1914 he was at the pinnacle of power. His phenomenally successful Tuskegee Institute, the Alabama vocational school he founded in 1881, grew with his prestige as an educator and spokesman for the black race. Washington, born a slave, preached a program of Negro accommodation and self-reliance—without integration. His message drew the praise and support of powerful whites from Boston to Montgomery, especially since he downplayed public demands for social equality. Washington told blacks not to waste energy complaining but to work hard within the system to help make America "the garden spot of the earth."[27] This approach made him acceptable to all but the severest white bigots. It also drew the criticism of far more radical and intellectual blacks like W. E. B. Du Bois, who accused Washington of blaming blacks for their own condition, and representing "in Negro thought the old attitude of adjustment and submission. . . . At a time when Negro civil rights called for organized and aggressive defense, he broke down that defense by advising acquiescence or at least no open agitation."[28]

Dissenters like Du Bois were relatively few and Washington, through his secret control of many black newspapers, contained their criticism. To whites in power during the early twentieth century, he was *the* black leader, a convenient singular spokesman and adviser for ten million Americans. President Theodore Roosevelt braved an enormous uproar by having Washington at a White House dinner in 1901. Young John D. Rockefeller, Jr., served him supper at his parents' house. Harvard University granted him an honorary degree. Even Queen Victoria had him to tea.[29]

Booker T. Washington arrived in Muskogee on the evening of August 18, 1914, in a style unlike any other African-American. Instead of a filthy segregated compartment behind the locomotive, he and his entourage of hundreds of sophisticated followers entered the city in a specially chartered train of well-appointed Pullman cars. A crowd of five hundred local blacks thronged the Muskogee rail depot. Members of the African-American press were also out in force. Editors and reporters representing four hundred black newspapers across the country covered the event—perhaps the greatest number of newsmen ever to have assembled in Oklahoma. Washington, broad and elegantly attired, stepped gracefully from his car and waved to the cheering crowd. Here was the embodiment of their future. Even to those without the right to vote, without legal protection, without money, without jobs, even to the most destitute among them, here was a man whose very presence evoked inspiration. To those trapped in an ugly caste system, Booker T. Washington was a minister of hope, hope for whites as well as blacks.

Oklahoma's black leaders had lobbied for Washington's visit for seven years. The purpose of his Muskogee appearance was to kick off the annual gathering of the National Negro Business League, an organization he founded in 1900 as a sort of black national chamber of commerce for farmers and businessmen struggling within a segregated economy. The three-day convention attracted more than three thousand League members from other states. Thousands more from around Oklahoma flocked to Muskogee to hear the legendary leader and witness the greatest display of black economic power ever assembled in the Southwest.

Muskogee's modern convention hall opened on the morning of August 19 for a series of Business League meetings. Located on South Second Street, down the block from the black business district, the large brick building was the pride of the city. It had been built in 1907 as a world-class convention site and appointed with fireproof plaster and nine hundred electric lamps generating "1,000,000 candle power lights." Although the hall seated five thousand, it proved too small to accommodate everyone wanting to hear Booker T. Washington.

After a day-long series of testimonials by successful black businessmen and farmers, Washington delivered his annual address. The night was hot and sultry, but men, women, and children waited for hours to get standing-room space. Some eight thousand blacks (including Jake Simmons and his entire family) jammed into the hall; five hundred more were unable to get in and stood outside. A good number of white citizens, seated in a special section, were also present (although most curious whites waited for a fully separate but equal speech which Washington gave two nights later). The moment Washington appeared on stage the audience gave him a standing ovation, applauding wildly and waving handkerchiefs in the air.[30]

Washington's discourse, as always, was filled with statistics and managed somehow to prove worthy of both the white segregationist's trust and the African-American's dignity:

> . . . I believe that the time has come when we as a race . . . should get off the defensive . . . and begin to inaugurate everywhere an aggressive and constructive progressive policy in business, industry, education, moral and religious life and in our conduct generally. . . . All the energy you have to "knock" with, all the energy you have to voice complaints, coin that energy into improved methods of handling your merchandise. And so with general race matters, damning the other fellow does not push us forward. His damning us cannot permanently hold us back. . . . Let us exalt the white man who treats us with justice and overlook and pity the little man who would retard our progress.
>
> . . . Of the 20,000 colored farmers in Oklahoma, 13,000 of them are without livestock and 3,300 are without poultry on their farms . . . While the Negro farmer is neglecting his opportunity of raising livestock, the prices are continually getting higher. Beef is being imported from Australia and from South America. . . . There is no special color line in stock and poultry raising. If the Negro has cattle for sale, they

will bring the same prices on the market as the white man's cattle will bring.[31]

The following day, Jake Simmons and other rural members of Oklahoma's National Negro Business League had a chance to demonstrate to Washington what they had done with life's opportunities. Although he owned as much land and more livestock than most of the successful farmers delivering testimonials, Simmons chose to show instead of tell his progress, using an afternoon parade of black agriculture and achievement to do so. The procession formed at the north section of town, marched down Muskogee's "Broadway," then turned south through the black business district on Second Street, past the convention hall where Washington and other dignitaries sat at a reviewing stand.

The parade began with a series of beautifully decorated floats from every major black church, civic, fraternal, and business group in the area. Some were covered with flowers and colorfully attired children, others with agricultural and industrial exhibits. A troop of boy cadets marched, resplendent in their tailored uniforms, and blacks rode proudly behind in their automobiles, carriages, and bicycles. Four bands kept up the marching rhythm, playing popular contemporary and religious tunes. Behind the largest of the bands two men on horseback held high a wide banner which read "Jake Simmons and His Cowboys."

Behind the banner rode the most spectacular array of black cowboys ever seen. The parade to honor Booker T. Washington had sent Simmons combing the countryside as part of the National Negro Business League, adding every available horse and "cowpuncher" to his troop. All his sons and some of his daughters rode with the group. They strode three horses abreast, line after line, enough horsemen to fill five companies of cavalry. The men wore jeans, leather chaps, shiny silver spurs, and colorful flannel shirts. Identical red bandannas hung from their necks, and atop five hundred heads sat white "five-gallon" John B. Stetson hats, their wide brims like sails in the warm breeze. The most attractive young women in the county, with complexions ranging from jet-black to caramel Creek "half-breed" color, drew the wildest applause of the group by dressing like cowboys in sequined rodeo attire. Spectators lined the streets and sidewalks; some climbed rooftops and trees. They cheered wildly and tossed trinkets and small gifts at the riders. John Simmons, riding with Jake Jr., remembers that the brothers felt immense pride showing off for Washington, whom they revered more than they did the President of the United States.

The parade lasted seventy minutes, with cowboys stopping from time to time to perform lasso tricks. Jake Simmons's group stole the show. The *Tulsa Star,* one of the state's largest black papers, reported, "Thursday afternoon, the best and longest industrial parade ever seen in the Southwest marched through the streets of Muskogee. . . . A very spectacular feature of the parade was the float of Jake Simmons of Haskell, Okla., followed by an imposing cavalry of five hundred Negro cowboys on horseback."[32]

Booker T. Washington had a policy of staying at the homes of local black families when he traveled. Two days after the parade Simmons brought a very unusual guest home for supper. Simmons had arranged for Washington to spend the night at his ranch. The family's immaculate house was even more spotless than usual. The guest room was readied, and Rose unlocked her cedar chest to break out the hand-embroidered linens and Gorham silver. Jake Jr. sat wide-eyed with his family at the dinner table. Although he generally loved speaking to adults, this time he just listened.

Washington talked about agriculture and black business for a while, but most of his conversation focused on Tuskegee Institute. After all, he was sitting with a prosperous landowner and three school-aged boys: fifteen-year-old Eugene (Jake Sr.'s son by Rose's sister), fifteen-year-old John, and thirteen-year-old Jake Jr. Dr. Washington was a promoter extraordinaire—not the sort to let an opportunity like this pass unnoticed. He urged Simmons to send his boys to Tuskegee, where they would be uniquely trained for a competitive society. Washington impressed his listeners, as he always did, with the accessibility of both his person and his ideas. "He didn't try to play dignitary," recalls John Simmons. "Everyone knew he was a down-to-earth person."[33]

Washington repeated a variety of central themes wherever he went, so his words to the Simmons family were probably similar to these from his autobiography: "When a Negro boy learns to groom horses, or to grow sweet potatoes, or to produce butter, or to build a house, or to be able to practice medicine, as well or better than some one else, they will be rewarded regardless of race or color. In the long run, the world is going to have the best, and any difference in race, religion, or previous history will not long keep the world from what it wants."[34]

Jake and Rose Simmons were not hard to convince. They doubted the local schools would be able to provide as good a secondary education, and had already been approached to send their children to Tuskegee by a local high school principal who had studied there. Jake Jr. spoke to every Tuskegee alumni he could find, and longed for its prestigious collegiate environment. Although he was technically half a year too young to attend Tuskegee, the principal probably made it clear to young Jake that his age posed no problem. By the time Washington left the Simmons home early the next morning he had secured an assurance from Jake Sr. that all three of his boys would be sent to Tuskegee.

Simmons was a man of action. He and Rose immediately began preparing their sons for their first trip out of the state. Just two weeks after Washington's visit the brothers, each with one trunk, boarded a segregated rail car in Haskell. Jake Jr., though two years younger than the others, was entrusted with holding everyone's school money and tickets. They arrived in Alabama late the next day after a sleepless, anxious night. On September 8, 1914, Jake Jr., Eugene, and John W. Simmons entered Tuskegee Institute, the center of black vocational education in America.[35]

3

LEARNING FROM THE WIZARD

There is no great education which one can get from books and costly apparatus that is equal to that which can be gotten from contact with great men and women.[1]
—Booker T. Washington

Tuskegee Institute rested like an oasis in eastern Alabama, the heartland of the racially ambivalent South. Booker T. Washington, the "Wizard of Tuskegee," strived to make the school all things to all people. To his white supremacist neighbors it was a plantation yielding a predictable harvest of obedient underlings. To the northern industrialists upon whom he relied for contributions it was an efficient factory producing a steady stream of hard-working, skilled laborers. And to fifteen hundred black students, who came from near and far to undergo the miraculous transformation from vassal to prosperous breadwinner, Tuskegee was a chapel of self-sufficiency and racial pride in a landscape of oppression and despair.

Washington ruled Tuskegee with the egalitarianism of a man who never forgot where he came from. After starting life as a slave he got through secondary school only because he was given a janitorial post to pay his way. Upon founding Tuskegee in 1881, he decreed that tuition would be free and that all "worthy" students would be given the opportunity to work off their room and board. He devised various programs for students of all ages, from basic remedial study to standard junior-college-level courses, and nobody who truly wanted to attend was turned away.[2]

Which is not to suggest that Tuskegee was a free ride. Manual labor was mandatory for all students. Washington's unabashed goal was to operate "a first-class industrial school rather than a second-class academic."[3] Every other day was spent out of the classroom in the agricultural or industrial division learning by doing. There was little room for esoteric analysis. A chemistry class examined local clay. In arithmetic students calculated the cost of plastering a recitation room. Teachers of agriculture assigned boys essays on preparing turnip fields, while girls wrote about the best way to cook greens. Industrial instructors graded a student's paper on technical grounds, then passed it to English teachers to mark for grammar.

Washington put his workers to good use. While Tuskegee reaped the benefits he wanted them to "learn to love work for its own sake." All labor, he taught, was created equal, and his belief that "there is as much dignity in tilling a field as in writing a poem" became a school creed. It was students who kept the institution's 2,300 acres of grounds beautifully groomed, and students who maintained the mechanical infrastructure of eighty-three buildings, just as it was students who originally built most of the buildings—from bricks produced at the school's own kiln. Students also experimented in the greenhouse and worked the school's fields, helping experts like George Washington Carver transform the area's tired soil into highly productive farmland. Most of the produce, meat, and dairy products served in the dining halls was grown or raised there, and surplus food was sold for cash. Girls, who made up a third of Tuskegee's enrollment, worked just as hard as the boys, cooking and cleaning dormitory kitchens, tailoring uniforms, and staffing the school's hospitals and offices.[4]

Hard work built Tuskegee and Washington believed that hard work, not intellectual posturing, would build success for his students. When educators like Harvard president Charles W. Eliot questioned the wisdom of placing vocation over academics Washington dismissed the criticism as the meaningless conjecture of ivy-towered book-worms. According to biographer Louis Harlan he "had a lifelong hatred of the abstractions and generalizations in the discourse of college-educated men."[5]

To Washington old-fashioned discipline was far more important.

Members of the two-hundred-person faculty were known to tremble at his approach, and incoming students were made to understand the militaristic order of the institute from day one. The boys' school was run by a "commandant," not a dean, and everyone received the principal's handbook of regulations upon registering. The first order was to report to the school tailor to purchase a dark blue uniform and military cap. The uniforms, which created a sense of equality as well as discipline, were worn at all times. Boys were placed into an ROTC-type organization containing a hierarchy of ranks, from private to captain. Attendance at regular chapel service was mandatory, and students were forbidden to leave the school grounds without permission. The girls' dormitories were completely off-limits to boys. They were protected by bright lights, fences, and a vigilant school police force. Boys and girls sat apart in classes, dining halls, chapels, and assemblies. Female students were zealously guarded by prudish women administrators who were so stodgy that it took Washington years to convince them to allow dance classes for young women (they argued that dancing would create "a serious menace to good order").[6]

Washington regarded his students as emissaries of black America. He believed appearances were crucial and was obsessed with improving personal hygiene. Baths at "stated periods" were mandatory, as was the possession of "serviceable underwear," brushes, shoe polish, and table napkins. Toothbrushes carried special significance. In *Up from Slavery* he wrote, "The gospel of the toothbrush is a part of our creed at Tuskegee. No student is permitted to remain who does not keep and use a toothbrush. . . . It has been interesting to note the effect that the use of the toothbrush has had in bringing about a higher degree of civilization among the students."[7]

During his five years at Tuskegee Jake Simmons, Jr., learned a lot more about the world than how to brush his teeth. He set out on the "A Prep" track, the most advanced academic (as opposed to industrial) degree program the school offered. He sailed through high school and junior college courses in subjects like English, business, math, and technical drawing. History, which became a lifelong passion, was his favorite topic. He was the sort of teenager who learned more out of class than in, taking full advantage of Tuskegee's unique extracurricular environment.[8] Booker T. Washington, at the height of his fame, was able to attract the most important figures of the day to speak at the Institute.[9] Simmons never missed a chance to hear entrepreneurs like Sears, Roebuck tycoon Julius Rosenwald describe the secrets of their success. As he got older few days passed in which he did not find himself repeating one of the principles he had heard at Tuskegee, like Teddy Roosevelt's admonition, "I wish to preach not the doctrine of ignoble ease, but the doctrine of the strenuous life."[10] Few of the great men of his day, Simmons realized, would have succeeded without developing their argumentative abilities. Selling, charming, motivating, and altering people all depended upon being able to talk a good game. Jake, like Tuskegee's principal, honed his persuasive powers into a formidable tool, and used it even while a student. He was the sort of well-spoken student instructors favored, and he knew how to bluff a very good game.[11]

College was not entirely academics at Tuskegee. Like everyone else Jake had to choose a vocational specialization, or find himself assigned one. He made a beeline for the machine shop, shunning the school's famous agricultural department. Automobiles were becoming accessible to the average American for the first time, and Jake wanted to be prepared for the mechanized future—a future far from squealing animals and grass which never stopped growing.

Jake's brother John had entirely different interests. He studied harness making and worked in the fields, assisting in planting and seed selection. John chose an agricultural program which required little class work. He opted for night school and an "industrial" rather than academic degree. His most memorable task was caring for Washington's horse ("that was my pride"), a large white gelding named Dexter.[12]

The one activity the two brothers shared a liking for was the spit and polish regimen of the student military association. During their first year John and Jake entered as corporals and advanced to the rank of sergeant. By their senior year the Simmonses both managed to earn the organization's coveted captain's bars. Another student who rose to the top of the organization was a Liberian named Earnest Jones, who became Jake's close friend. Jones put his military training to good use: he eventually became Liberia's secretary of war—and an important contact in Simmons's future.[13]

Jake's association with African classmates like Jones proved helpful even during his school years. Booker T. Washington gave a special monthly lecture just for foreign students (most of whom were there on special scholarships). Jake used his friendship with both Jones and his first girlfriend at Tuskegee, who was from the Gold Coast (now Ghana), to finagle his way into the lectures. Here he finally came face-to-face with the man who was to be his greatest role model. Yet even the illustrious educator failed to intimidate him. Simmons later told his son Donald that he once approached Washington privately, after a lecture, and asked him point-blank the question which troubled many politically conscious blacks of the day: why did he avoid open criticism of America's degrading racial caste system? Washington, who had come to appreciate the forceful young Oklahoman, candidly replied, "This is what I have to do to have a school like ours in the middle of the South."[14]

By the early twentieth century a growing number of African-American activists felt that Washington's words rang hollow. They thought that his famous endorsement of segregation was an opportunistic sellout. Washington had publicly stated, "In all things that are purely social we can be as separate as the fingers, yet one as the hand in all things essential to mutual progress."[15] Although he secretly financed legal battles to overturn Jim Crow laws and allow blacks the right to vote and serve on juries, he refused to fight openly for such issues. In 1909 a coalition of Washington's critics formed the National Association for the Advancement of Colored People (NAACP). They were determined to rescue the black struggle from the accommodationist conservatism of Tuskegee's principal. Their integrationist activities were so controversial that when word leaked about an early association dinner at which white women would be sharing tables with black men, it evoked scandalous headlines in every paper in town—and this was in New York.[16]

Washington told his black detractors much of what he told Jake Simmons: in his part of the country racist opponents did a lot more than write nasty editorials. If he openly fought segregation, Tuskegee Institute, and probably his life, would be crushed. He believed that nonsoutherners lacked an understanding of the hateful tide he swam against. This was, in fact, far more than paranoia. Although he regularly tried to placate locals by referring to white citizens of Alabama and Tuskegee as the Negro's very best friends, his graciousness failed to mask the success of his work, and it was this success that he was resented for. Tuskegee's whites had only a few decaying plantations to be proud of. These were no match for the stately brick structures of an institution endowed with $2 million. Alabama politicians often curried voter favor by denouncing the school and questioning its financial affairs. The state legislature was a forum for anti-Tuskegee debate. A representative there noted, "I believe Booker Washington and his gang would prove to be the curse of the South, and if I had my way I would wipe his institute off the face of the earth."[17]

In light of Washington's power more demands were made than threats. An editorial in the *Mobile Herald* charged, "If it be claimed that the whites are benefited by educating the Negroes, then let us see some cooks come into white kitchens from Tuskegee schools."[18] Even Washington's fanatically utilitarian approach to education did not deter criticism that he was turning good fieldhands into worthless dreamers. In 1903 a white lawyer from Montgomery named Gordon Macdonald jumped on the anti-Tuskegee bandwagon and was quoted in newspapers across the country berating the school. Although he had never been there, Macdonald observed that "for one genuine, hard-working husbandman, or artizan [sic] sent into the world by Washington's school, it afflicts this state with twenty soft-handed Negro dudes and loafers, who earn a precarious living by craps or petit larceny, or live on the hard-earned wages of cooks and washerwomen whose affections they have been able to ensnare." Tuskegee's girls, added Macdonald, were taught to shun hard work "while their poor mothers toil over the wash tubs and cook stoves that their daughters may be taught music and painting . . . and to rustle in fine dresses in a miserable imitation of fine ladies."[19]

A few years after Simmons graduated, a mob of local white racists attempted to put into action the rhetorical opposition harbored by many Alabamans. In July 1923, during the tenure of Tuskegee's second principal, Robert R. Moton, the barbarians came close to breaching Tuskegee's walls. Moton had defied a group of Ku Klux Klansmen who threatened to kill him and blow up the Institute if he did not relent on his insistence that a new federal hospital for black veterans be staffed by black doctors and nurses. The night before Independence Day, a small army of hooded Klansmen burned a forty-foot cross in the town of Tuskegee, then piled into a seventy-car caravan and headed toward the campus to raze it. The Klansmen were shocked to find that Moton had summoned help from Tuskegee alumni across the state. These reinforcements augmented the student military association, and the defenders strategically positioned themselves, with rifles, around buildings and access roads. When they realized what they were up against the Klan backed down. The six-hundred-bed hospital was soon opened, and run by a staff of black professionals.[20]

By the time Jake returned to Oklahoma from his first year at Tuskegee he was a lot more like the man he had always claimed to be. Yet to his dismay his father still treated him as a boy. The elder Simmons met his sons at thc station in Haskell as they returned home for summer break. They showed off their uniforms, climbed into the wagon, and began telling him about Tuskegee life. On the way to the ranch, Jake Sr. stopped at a store and bought each a pair of overalls. "Put them on," he said. "There's plenty of daylight left and you boys are going out to the field to work." This was not the welcome Jake Jr. expected. "What about seeing Mama?" he protested. His father would not hear of it. "You'll see her after dark when y'all finish chopping cotton," he said. That afternoon, sweating in the cotton field, Jake made up his mind never to spend another summer in Oklahoma.[21]

There was one saving grace to summer recess: Willie Eva Flowers. Jake had kept in touch with the mixed-blood black Creek girl since meeting her at the Tullahassee school a few years earlier. Eva was beautiful and mature, with a quiet, cautious "Creek manner" similar to Jake's mother, Rose. Simmons charmed her effortlessly, with the natural ease of one sure of fulfilling his objectives. His bold aspirations and tales of the great men he met at Tuskegee set him apart from other boys. Before the summer was over Jake had talked his way into her heart.

In the face of Jake's self-promotion farming was an embarrassment. Overalls symbolized the limitations of a rural livelihood: no self-respecting businessman entertained guests in overalls. Whenever Eva was near he made sure to dress in his school uniform or a well-pressed jacket and tie. Once, when she arrived unexpectedly at his father's ranch, he hid out in the fields for an entire day because he did not want to be seen wearing overalls. Impressions, he had learned from Booker T. Washington, were important, and he was determined to leave his hometown sweetheart with the image of a man who was going places.[22]

Simmons's second year at Tuskegee was marked by the death of Booker T. Washington. Despite periodic overexhaustion Washington never paused from his demanding schedule. To him the meaning of life was found in the process of work. He labored till he dropped, at the age of fifty-nine, in November 1915.[23]

Jake Simmons never tired of quoting his old principal, especially when lecturing youngsters. Washington's greatest skill—his oratory technique—became an essential part of Simmons's repertoire as a businessman and activist. Nothing, Washington believed, could take the place of the "soul" of an address. He advised speakers to first have something important to talk about, then forget about grammar and sermonizing and provide their audiences with succulent, irrefutable facts.

Whether at a political barbecue or in a corporate boardroom Simmons followed this advice, always driving his arguments home with a heavy dose of statistics and anecdotes. Like Washington, his statistics were not always absolutely precise. The important thing was that they seemed accurate, that they promoted a credulity in the listener, and a willingness to believe the arguments that accompanied them.[24]

During the summer of 1916 Jake Simmons stepped out of Tuskegee's protective enclave and entered the real world. Having vowed never to labor on his father's farm again, he looked for work in Alabama and landed a seasonal job as a Pullman porter. His neat appearance and poise helped him win the job, but he was deficient in one essential qualification: subservience. Simmons loved wearing the elegant porter's uniform but he never made it past his maiden voyage. Soon after the train got moving a white passenger yelled, "Boy, come here and get my bags!" Simmons got the bags, then politely protested that he was not a boy and resented being called one. The passenger looked at him strangely, then said, "Young man, I hope you take this advice the way I mean to give it to you because it's for your own good. If you don't want to be called a boy in life then don't do a boy's work. Because boys carry bags for men." Simmons thanked him, and when the train returned to its starting point he turned in his uniform and walked off the job.[25]

The word "boy" was like a razor to Jake Simmons's pride. It implied that the color of his skin made him unfit to be called a man, no matter how well he dressed, how articulately he spoke, or how politely he acted. He swore never to do a "boy's" job again.

Men's work, Simmons decided, could be found in the enlightened North. News had spread across the country of Henry Ford's lucrative five-dollar workday and workers from all over flocked to Detroit to profit from the Model-T revolution. Although Henry Ford was a virulent anti-Semite (Adolf Hitler hung the businessman's picture on his wall), when it came to giving blacks a place in his factories he was one of the most progressive industrialists in the country. Robert Lacey, in his recent biography *Ford,* observed that in the mid-twenties, one-tenth of the automaker's 100,000-person work force was black, and Henry Ford "never paid a man more, or less, on account of his skin."[26]

When fifteen-year-old Jake Simmons arrived at Ford's employment office he saw a long line of unskilled whites waiting to apply for work. He went directly to the man in charge of hiring and said he was a trained machinist.

"How can you be a machinist?" the Ford man asked skeptically.

"I study at Tuskegee Institute," Jake replied.

"Oh," said the man, suddenly understanding. "Where Doctor Washington is."

Simmons was hired on the spot. It was a firsthand affirmation of Booker T. Washington's belief that merit could transcend the color barrier. Donald Simmons explains, "It left the impression on him that you could even leave white people standing in line if you were qualified to do something that no one else could do."[27]

Jake entered his third year at Tuskegee with a clearer sense of his place in the world. He took business-related courses and moved into the prestigious Rockefeller Hall. The fact that the hall, the largest building on campus, was reserved for seniors made it an even more attractive home for Simmons, who talked his way into a room.[28]

Jake wore his collegiate good looks with the worldly expression of one who had been around awhile. He became a popular figure on campus, playing on both the football and basketball teams. He is easy to spot in a photo of Tuskegee's 1919 basketball team. He sits in the center of eight teammates, holding a basketball on which "Champions 1918" is painted. While the rest of the players wear sleeveless athletic shirts, Simmons, about six feet tall and lanky, sports a trendy long-sleeved sweater. The other boys all lean in toward him; his cavalier expression conveys a sense of leadership and sophistication years beyond that of his costudents.

Jake's two brothers had little interest in becoming leaders. John was a team player; one of his proudest memories of Tuskegee is playing trumpet in the school's 225-man band. "Tuskegee had the largest band in the South," he says. "And I was the second trumpet player—a country boy! What do you think about that?"

Jake's half-brother Eugene didn't bother with extracurricular activities. After a few years at Tuskegee, he didn't even bother with school activities. In the crapshoot of tribal land allotments Eugene had been dealt the jackpot: oil-soaked acreage near the famous gusher at Glenpool. By the time he turned eighteen, oil wells on his property were yielding $500 a month in royalties—a small fortune in those days. Once the checks started coming directly to him, Eugene dropped out of school. While his father thought he was living in a modest dormitory he became a local playboy, keeping a fine house in town, two automobiles, and more than one girlfriend. School authorities lost track of him. Jake Jr. and John realized what was going on as soon as they saw Eugene parading around town in fashionable "civilian" clothes (students always had to wear their uniforms), but an occasional cash gift of twenty dollars overpowered any urge to snitch. Eugene's oil wells kept pumping. He never went back to school. According to John Simmons he spent the rest of his life as a "roustabout and an alcoholic."[29]

Although Jake Jr.'s lifestyle was never as flamboyant, he enjoyed the company of women as much as any of his brothers, and was at least as charming. Given the strict separation of the sexes at Tuskegee this posed a considerable problem. But rules, like most obstacles, bent under Simmons's will. By the time he was sixteen he had learned the ropes at Tuskegee and there was no stopping him. He moved freely on and off campus at night, evading the watchman at the main gate by sneaking through the delivery passage used by supply wagons. When he wanted to get a message to a girl during an assembly he would pass a note through dozens of classmates seated between them. Before long Simmons had become a discipline problem to Tuskegee's staff. "He almost got put out for kissing a girl in the hallway," explains Donald Simmons, "He was always up to some kind of prank. And, of course, with the strict rules at Tuskegee in those days they might expel you for looking at a girl cross-eyed. He told me he would have been put out two or three times if it hadn't been for Mrs. Washington. She was really Daddy's mentor down there."[30]

Margaret Murray Washington was Tuskegee's Eleanor Roosevelt. Her influence lingered well beyond her husband's death. Known as the "lady principal," Margaret headed the school's Department of Women's Industries and was a key member of the executive council which ran Tuskegee. During the leadership vacuum following her husband's death, she was instrumental in creating a new administration. Margaret was a portly, dignified woman whose frameless spectacles added to her image as a wise old matron. Well-liked and trusted, hers was the office of last resort for students in trouble.[31]

Jake wound up there many times. His Tuskegee friends later told his oldest son J. J. III that "he was always involved in whatever devilment was going on." Good grades and Margaret Washington's intervention allowed Simmons to get away with almost anything. He cavorted with different girls and then, during his final years at the institute, he settled down with a steady named Melba Dorsey.

Jake was intrigued by people with backgrounds different from his own. Melba was from Chicago, a northern metropolis. Her parents, far from living a traditional life as laborers or farmers, were both artistes. Her mother was a singer who often toured the country with well-known choirs. When she did not bring her children on the road they were left with their grandmother. Melba's father also traveled frequently, costumed in ornate knickers. He was a daredevil tightrope walker; Melba describes him as "an acrobat who did wire work." When she was little her father introduced Melba to show business by putting her on a chair during his act and carrying her across the tightrope. It was the most frightening experience of her life.[32]

Melba Dorsey resembled Jake's hometown sweetheart, Eva. Both were medium height, with shapely figures by today's standards (considered slightly skinny then). Both dressed well and carried themselves with a refinement which distinguished them from most girls. Melba also possessed an unconventional glamour. She was never the homemaker Eva would prove to be. After attending a girls' finishing school in Chicago she had come to Tuskegee to round out her education. Like her mother she favored singing, but given the school's vocational bent she spent most of her time studying typing and other office skills.

Despite her sophistication Melba was a bit shy. Simmons overwhelmed her. Since boys were not allowed to take girls out he regularly sneaked off campus, purchased special foods in town, and left it in secret spots for her to pick up. By the time he was an upperclassman the indefatigable "Mr. Jake," as Melba remembers him, "was very assured he was going to walk down the hall with me when we were changing classes, regardless of what the rules were or what floor we were on." There was no stopping him. Once, Melba recalls, her voice mixed with shock and admiration, he impersonated an instructor by putting on his civilian clothes and a man's derby and walked right across campus to the girls' dormitory, just because he felt like saying hello. "He frightened me," she says. "I didn't want him to do things of that sort because I didn't want any difficulties in my life."

Difficulties, nonetheless, soon followed. That Melba was the object of his troublemaking was more than a little flattering to the sixteen-year-old. She became increasingly enamored of the audacious Oklahoman, even after his best friend revealed that he had a girlfriend back home. Jake was her first love, and she fell hard. Even today, at the age of eighty-seven, Melba turns wistful when she speaks of him. "We met in the dining room every day," she says. "We visited each other's table. We shouldn't have, but we did. He was very aggressive, you know. What he wanted he wanted and he simply went for what he wanted."

Simmons's less social moments as a Tuskegee upperclassman were spent pursuing the two subjects which interested him most: Negro history and public speaking. He spent weeks researching and writing contemporary speeches about the struggle for African-American dignity. Obsessed with the injustice of racism, his conviction provided the soul which Booker T. Washington had said belonged in any worthwhile public address. Standing at the pulpit of the Institute's empty chapel he rehearsed his delivery again and again, practicing the pitch and intensity of his speeches. "Boy, you could just hear him all over the school," Melba Dorsey Wheatley recalls. "I could hear him talking away just sitting in my dormitory."

Simmons spoke regularly at student gatherings and became one of Tuskegee's most popular orators. An early discourse called "The American Negro as a Soldier" won a prize in a First Boston Trinity College Church competition. Delivered when he was seventeen, it discussed the war record of African-Americans through the years, then noted,

> It seems to be an anomaly of fate that the American Negro, the man of all men who is held in disdain, should stand out in conspicuous relief at every crisis of our national history. His blood offering is not for himself or for his race, but for his country. . . . Any other class of people living under conditions to which the American Negro has been subjected would imitate Job's distracted wife and curse the white God and die; but the Negro will neither curse nor die, but smile and say to his country, "Though you slay me, yet I served you . . ."
>
> Friends, teachers, and schoolmates, think of the American Negro. Measure him not by the height which he has attained, but by the depth from which he has come, and I know you will agree with me when I say, he has come up to the standard as a soldier. . . . God will change the map of the world. He has seen that the American Negro will play an important part in bringing freedom to millions of people who are making the fight. May we hope . . . that the American black soldier will win in this war not only for America, but for full and unquestioned citizenship for his race.[33]

Simmons's hope was born of the unique sense of empowerment Tuskegee allowed its students. In their impressive, well-built enclave anything seemed possible. While the real America had lynched 264 blacks during World War I alone, Jake Simmons was telling his classmates about the brave new spirit of brotherhood sure to arise from the great conflict. In another speech he backed the radical new idea of a League of Nations, writing, "The League is needed to express the increasing democracy and solidarity of the human race. The cooperation of the Allies has been the world's most successful experiment in brotherhood. The serious problems that now confront the world cannot be solved by the nations acting separately."[34]

Most of Simmons's speeches focused on black pride. Their tone was considerably more militant than Booker T. Washington's, and far better suited to the NAACP's new era of direct confrontation. He evoked racial pride by excavating little-known facts about black history—especially African history. In one speech he glorified the ancient Nubian capital of Meroe and the city of Thebes, noting:

> lp;&6qThese two Negro cities were so civilized and prosperous that Rome and Greece stood transfixed before them. Egypt borrowed her light from the venerable Negro up the Nile; Greece went to school to the Egyptians; Rome turned to Greece for law and science of warfare; and England dug down into Greece twenty centuries later to build, plant, and establish a government. Now, who gave the world its law, its science, and its government? The black man of ancient times.
>
> . . . Centuries have flown apace, tribes have perished, cities have risen and fallen and even empires, whose boast was their duration, have crumbled, while the two Negro cities, Thebes and Meroe, have stood. Notwithstanding their degradation . . . this remarkable and enduring Negro race has not been blotted off the earth. They still live, and are increasing. Certainly they have been preserved for some wise purpose to be unfolded in the future. . . . Who can tell that this race that once ruled the world may yet again in God's good time, stretch forth her hand of power and greet this world again with a new inspiration, a long and successful reign.[35]

Simmons, not surprisingly, was elected vice president of his senior class and its key graduation speaker. Graduation was an important ceremony at Tuskegee. Out of fifteen hundred students fewer than one hundred graduated each year. Most never received diplomas. For those who did, observed historian Louis Harlan, graduation "was almost an ordination as a minister of the Tuskegee gospel."[36] The class hoped that Simmons could create a sermon expressing their collective faith. He did not fail them. On the evening of May 21, 1919, as hundreds of commencement visitors assembled on the "White Hall Lawn" of the campus, Jake delivered the opening address. Sensing the weight of history behind him Simmons spoke of a need to struggle and articulated the hope of his generation, a generation that would later transcend fear and lay the groundwork of the modern civil rights movement. The grade's "class creed" would be the guiding doctrine of his life:

> We believe in our motto, "No Victory Without Labor." We do not believe that we, as a class, could have selected any more inspiring words than these, had we searched through the entire vocabulary of the world, with all its many languages and dialects. Everybody, whether he knows it or not, has one predominating principle in his life and one supreme influence and one inspiring force by which all his acts are arranged and shaped. If this principle is a worthy one, his life will be one of nobility and honor; if it happens to be an unworthy one, his whole character will descend to the depth of wreck and ruin. It therefore stands us all well in hand to think strongly, and seriously of the principle that is to be the foundation of our life pilgrimage toward the perfection we secretly, if not openly, desire to attain.[37]

Melba Dorsey watched proudly as the crowd applauded. She would never graduate from Tuskegee. After Jake picked up his diploma the couple moved north and were married in the Chicago home of her mother. Melba's friends and relatives attended, but Jake Sr. and Rose did not even know about the wedding until weeks later.[38] A daughter was born to Melba Dorsey in March of the following year.

It is impossible to know whether Jake intended to spend the rest of his life with Melba. He never actually lived with her in Chicago. Right after the wedding he returned to Detroit, where he had worked the previous summer at the Packard Motor Car Company. Why he didn't go to a university at this point is uncertain; he once told a friend he had a scholarship offer from a good school but didn't take it.[39] It might have been family obligations, or even a belief, like Booker T. Washington's, that work experience was more important than book knowledge. Simmons was ready to stake out a career, and he imagined a big northern factory was just the place to start. It would be the only time in his adult life that he ever worked for anyone.

Jake dreamed of using his machinist skills to become an inventor. After a few months building cars at Packard he drew up plans for a windshield defroster which worked off the engine exhaust. He got a patent for it and tried to interest his superiors, but nobody paid any attention. They did not think a black man capable of inventing anything useful. "Because of the discrimination that he witnessed after finishing Tuskegee and going to Packard," his son J.J. III explains, "he said he realized he wasn't cut out to work for anybody."

Simmons wanted to go into business for himself. But he had few opportunities or contacts in the North. Furthermore, his father had found out about his secret marriage and was infuriated. While making a trip to the Kellogg sanitarium in Battle Creek, Michigan (to receive medical treatment for spinal meningitis), Jake Sr. stopped in Chicago to see what his son had gotten himself into. "He came to straighten out affairs," Melba Dorsey Wheatley explains with intentional ambiguity. "He was very surprised to see the kind of people we were and the way we lived. I think sometimes people used to think that girls who were born in the city were kind of fast or something of that sort, you know—looking out for themselves for whatever somebody had."

Melba and Jake's daughter, Blanche Jamierson, having spent a portion of her childhood in both Chicago and Muskogee, takes a more geographical view. "This was a class thing," she says. "There was a difference in the attitudes of people in the northern and southern states. They thought the women, especially in Chicago, were just fast women. And I presume they warned their sons and daughters about these fast northern people and their loose lives."

Jake Sr. strongly believed in what he called "pedigree," and family trees. The rancher regularly lectured his children and grandchildren that "people take more pain breeding cows and horses and hogs than in breeding their own children."[40] According to Melba, Simmons left his son with an ultimatum: get divorced or get disowned. "He was stubborn just like Jake [Jr.] was," Melba says. "Some people choose a husband or wife for their children and they expect it to be that way. I know he had Eva in mind, and she was a lovely person. He was ready to back Jake in what he wanted to do [career-wise] if Jake did what he wanted him to do [get divorced], and unless he did that he would have disinherited him. Jake [Jr.] was terribly upset. I felt he was torn and didn't know what to do. A career was standing there for him and we don't always have those things open to us."[41]

Without Jake Simmons, Jr., alive to say why he decided to leave Melba Dorsey it is not possible to explain fully his decision. In addition to his father's ultimatum there were other considerations. Jake Sr. had brought promising news with him. Oil had been found on the land adjoining his son's 160-acre allotment in Muskogee County. To a would-be oilman, this made Jake Jr.'s Oklahoma property the perfect place to start an independent career. Furthermore, Simmons's experience with Packard had taught him that the only way to be treated with dignity at work was to become his own boss. Between his land allotment and the booming oil industry Oklahoma offered far more opportunity than Detroit or Chicago. As Simmons's northern prospects began to look less and less appealing, so did his rushed marriage.[42]

Just before Christmas in 1920, Simmons traveled to Oberlin College in Ohio, where Eva was studying, and married her at the county courthouse. Eva became a loving mother to Blanche as well as the three sons she had with Jake. Jake paid child support from the moment he left the North.[43] Melba is quick to point out that Jake always maintained good relations with her and took exceptional care of their daughter. "He was always there for her," she says. "When I started on my road work [singing with a group as a "jubilee trouper"], half the time I would come home and she was gone in Oklahoma because he would come or send for her. He never deserted his own."

As for her personal feelings, Melba says that the past seventy years have helped smooth over her disappointment. "I have no enmity," she says calmly. "That's foolish. As you get older you learn not to grieve over things any more than is necessary. You don't have time for grieving so much. Especially when you let God take charge of your life, you have no bitterness, you only have love. You know whatever happens has happened . . . and bearing crosses one becomes stronger."

Jake and Eva would have their own crosses to bear. But their marriage would last until the day he died, and create the domestic foundation upon which Simmons could build a legend. He had strayed from Oklahoma, strayed from the sweetheart of his youth, and strayed from his destiny. But now he was back on the frontier. The most famous of all of Booker T. Washington's tidings had shone through a troubling time: "Cast down your bucket where you are," said Washington, urging black southerners to stay put and "bear in mind that whatever other sins the South may be called to bear, when it comes to business, pure and simple, it is in the South that the Negro is given a man's chance in the commercial world."[44]

Another chance was what Simmons wanted. He quickly headed toward Oklahoma, eager to stake out the sort of destiny which the Wizard of Tuskegee had taught him to thirst for.

4

THE LEASE HOUND

There were all kinds of skulduggery going on. Black landowners didn't trust the white man, and damn well plenty reason not to. . . . most of them . . . wanted to screw up the leases and not give them a dollar.[1]
—Taft Welch, senior chairman, Western National Bank, Tulsa

Jake Simmons, Jr., was part of a new generation, one of millions across America abandoning their parents' rural farms for the fast-growing cities. He returned to Oklahoma, but refused to live the life his father envisioned for him. Jake Sr., past his prime, was often ill, and he wanted one of his sons to help oversee the ranch. Yet in spite of their upbringing not a single one of his seven boys was willing to commit himself to the land. "He didn't have one son who was a good farmer or rancher," explains Erma Threat, his granddaughter.[2]

Jake Simmons, Jr. had other ideas. He returned to Oklahoma at the end of 1920 with his new wife, Eva Flowers, and moved into the city of Muskogee. It was near enough to enjoy the benefits of Simmons's family, yet far enough to establish independence.

Muskogee lies at the crossroads of the Arkansas and Grand rivers. Unlike most of the dry, scrub countryside of Oklahoma, its proximity to water makes it relatively green and bountiful. The city's wide avenues are lined with shady elms, and its downtown architecture embodies a frontier sensibility. In 1920 its 30,000 residents made it Oklahoma's third largest city, although the urban sin of modern municipalities like Oklahoma City and Tulsa were—and still are—noticeably absent. Even today magazines like *Playboy* are virtually banned, and churches outnumber places serving alcohol at least ten to one. By 1920 the staid middle-American values described in Merle Haggard's famous country music song "Okie from Muskogee" were already evident: "We don't make a party out of loving,/ But we like holding hands and pitching woo!/ We don't let our hair grow long and shaggy/ Like the hippies out in San Francisco do."[3]

The Simmonses built their house at 2808 Court Street, a few miles from the city center, on land which Eva's grandparents bought with oil royalties from her tribal allotment. It was a long ranch-style house of flat tan stone, with a spacious lawn. Although sections of Muskogee were strictly segregated, the neighborhood around west Court Street was the largest residential district. It was also uniquely well integrated. There were few incidents of racial hostility. Children of different races did not attend the same schools or churches but many played together. To this day on summer afternoons when many residents sit outside a visitor can observe porch after porch of racially mixed neighbors fraternizing to an extent inconceivable in many American cities.

Although Simmons never lived on his 160-acre tribal allotment in Wagoner County the land gave him a chance to learn about the oil business. His childhood ambition was to become a driller like the Tulsa tycoons regularly glorified in local newspapers. But drilling required big money. Even the simplest wells cost thousands of dollars, and nobody could say for sure where oil would be found. Drilling was a rich man's gamble, and Simmons was far from rich. Most farmers leased their land to an oil company or broker. In return the company financed the well and gave the landowner option money up front, plus a royalty percentage, generally one-eighth of all the oil found.

Simmons got to learn about brokering by wheeling and dealing for his own account. Because oil had been found on a neighbor's property, drillers were eager to purchase leases to explore his land. He was therefore able to divide the land into small pieces and make different deals—and contacts—for each. He befriended an older broker and quizzed him every time he had a question about the business. "He liked to be with people he could learn from," observes Simmons's niece, Erma Threat. "He aligned himself with people who knew what he wanted to know."[4]

Oil was never discovered on the land, but it was a valuable experience. He had found his way into a profession that needed no start-up money, maximized his persuasive skills, and still allowed a direct share in the tremendous potential of an oil well. Simmons became a broker.

It was a wise decision. Blacks owned considerable land in the area, and were as eager to profit from their potential oil reserves as anyone. His well-known family name helped gain entry to the county's Creek and black communities, especially in the Haskell area near his father's house. Jake also brokered a few profitable wells on his parents' ranch. He made money in oil, but to ensure a better living he expanded his activity to include a general real estate practice. He opened an office in downtown Muskogee, joining dozens of black companies which thrived there. (So successful were the city's black merchants that Stetson hats and Justin boots both selected a black store as their exclusive franchisee.)[5]

Simmons's real estate agency helped people profit from their land, whether it meant buying or selling farms or optioning off mineral rights. Sometimes he invested his own money to piece together leases in a specific area. By assembling such a block he was in a better negotiating position when it came time to flip the leases to an oil company with the resources to drill. Occasionally he kept a small percentage—5 percent or so—of the action for himself as an "override royalty." At other times he was hired by oil companies beforehand to pursue leases in designated areas. He handled most of the legal work himself, preparing leases and deeds, researching ownership titles, and registering transactions at the county clerk's office. He carried a portfolio of would-be deals wherever he went, ready to swap or sell any of a dozen leases at a moment's notice. He lived up to the term oilmen used in describing the service brokers like him provided: he was a "lease hound," and one of the best.

Starting out was far from easy. Although Simmons earned enough to invest in other people's oil wells, it took him a long time to acquire capital. He was never a particularly lucky man, and his father was rarely in a position to lend anyone cash. Jake Sr. suffered from chronic spinal meningitis, a condition which required periodic visits to the expensive Kellogg Clinic in Battle Creek, Michigan, for surgery and a curative regimen. The months-long cure cost the rancher tens of thousands of dollars. (Once when he did not have time to get to a bank

Rose shocked everyone by lending him $5,000 in gold coins which she had secretly hoarded in her cedar chest.) Although he was successful, Jake Sr.'s assets were always tied up in land. Whatever money he did not use to buy property or cattle was spent on his illness.[6]

To make matters worse, Eva Simmons's sizable oil income dried up soon after the couple were married. Her 160-acre tribal land allotment was near the Cushing oil fields, some of the most productive in the world. Five successful wells were drilled on Eva's land, producing nearly eight hundred barrels per day. Her one-eighth royalty paid thousands of dollars a month—but never to her.

Eva lost her oil income as a result of her guardian's mismanagement. Her mother had died during childbirth, and her father was sent to prison while she was young after shooting up a Muskogee streetcar because the driver threw him off for refusing to sit in the back. (The segregation laws of the new state did not sit well with part-black Creeks like Flowers.) Grandparents Annie and John Escoe were left to raise her and act as Eva's legal guardians until she turned twenty-one. They used some of her oil revenue to send her to Ohio, where Eva attended high school and then Oberlin College (as a teenager she even had enough money to buy a car). Although she left before graduating her formal education surpassed most women of her day, and if given the chance she would probably have been more successful handling her finances than her grandfather. John Escoe, known as a "high liver," did a better job spending Eva's money than managing it. He left such business to a team of white attorneys. By the time she became an adult Eva was shocked to find that her lawyers had somehow come to own most of what she assumed was hers.

Over the years John Escoe had been duped into misappropriating the proceeds of Eva's royalties to her attorneys in return for relatively small sums of money, which he promptly spent. By the time she turned twenty-one her royalty interest was nearly worthless. Jake and Eva did not even try taking her case to court. At this time hundreds of Creeks, many of them black orphans, were being bilked out of their mineral rights by corrupt lawyers. In case after case juries refused to indict the white swindlers. Furthermore, says Donald Simmons, his mother refused to sue because "For her to file against the crooked lawyers would have implicated her grandfather—she wouldn't send her grandfather to jail."[7]

Other domestic problems also hounded the newlyweds. Nearly a year after they were married their first baby died within three days of its birth. The tragedy did little to relieve the couple's anxiety over Jake's short-lived first marriage. Although their second child, Jake III, was born in 1925 with few problems, a third infant, Billy, was ill from birth, and died well before his first birthday.[8]

It was a difficult period, made harder by the most bitter surge of racial violence in Oklahoma's history. Within a year of Simmons's return from the north the area witnessed the bloodiest race riot in American history.

The Tulsa race riot had been brewing for a while. Few towns in the Wild West remained as wild as Tulsa, whose population quintupled between 1910 and 1921, to nearly 100,000 people. What Muskogee lacked in sin was more than compensated for by Tulsa's boomtown bounty of prostitution, gambling dens, drugs, and speakeasies. It was a lynch-happy mob town, where "whipping parties," encouraged by local newspapers, attacked an average of one African-American a day.[9]

Lynchings were so bad that when a contingent of black soldiers boarded trains to fight in World War I, they unfurled banners reading "Do Not Lynch Our Relatives While We Are Gone." To defend themselves, areas like Muskogee became centers of African-American militancy.[10] A black newspaper noted that lynch mobs had an uncanny respect for armed law-abiding citizens "who acted in earnest and on time."[11] In the years preceding the riot a number of white mobs were successfully dissuaded by such large groups, and in Tulsa a militant African-American society called for paramilitary squads to safeguard the community.[12]

On May 30, 1921, a minor incident between a bootblack and a young white woman sparked Oklahoma's racial powder keg. The man was arrested. Although the woman was never actually hurt the *Tulsa Tribune* rushed into print with a scathing fictitious article and a headline which reportedly read "To Lynch Negro Tonight." To hundreds of whites who mobbed the courthouse, the headline was like a party invitation. When carloads of blacks came to the rescue the riot broke out.

The white mob brought the battle across the railroad tracks into

North Tulsa, home of 11,000 African-Americans. They burned it to the ground (supposedly with the help of homemade bombs dropped from two airplanes), but not before suffering heavy casualties. By the time the National Guard brought the situation under control somewhere between thirty-six and 175 Tulsans of both races had been killed.

Whites knew they had carried their bigotry too far. Tulsa County never experienced another lynching. Black Oklahomans like Simmons saw the riot as a tragic but necessary affirmation of black power—a demonstration that their race was willing to go to war for its dignity. Donald Simmons notes that the ferocity of the black response differed sharply from the helplessness of African-American riot victims in other cities at the time. "You had a western heritage here—that you protected your own," Simmons says. "There wasn't a black family in this state that didn't have guns. So they just didn't roll over and let whites run crazy on them."[13]

Although the violence of the Ku Klux Klan decreased after the Tulsa race riot, its membership continued to grow. By 1923, in a state of scarcely two million, 103,000 Oklahomans were Klan members, including most of the state's prominent businessmen and politicians. For a while it controlled the state legislature, and its influence on the business community was felt for decades to come.[14]

Nonetheless, Simmons managed to find white oilmen willing to deal with him. He was able to provide his family with a life of middle-class comfort and even have something left over. After a particularly lucrative deal he took his savings and purchased the best car money could buy: a brand-new 1929 Pierce-Arrow. He and Eva drove it across the South to Tuskegee Institute's tenth-anniversary reunion for the Class of 1919. It was the status symbol of his success. At the festive reunion, Jake posed beside the Pierce-Arrow with his beautiful, well-dressed wife for a photo in the alumni newspaper. To this day Milton Nelson, a retired Chicago postal carrier who graduated from Tuskegee in 1921, recalls the photo. "Jake was really getting up in wealth by that time," Nelson says. "I remember seeing his picture with his wife and an automobile in the paper—it had all of us jealous, because we couldn't even get taxi fare."[15]

Soon Simmons wished he had never bought the expensive car.

Like millions of Americans the young broker saw 1929 as the finale of the roaring twenties, a harbinger of an even more prosperous decade to come. And like millions of Americans when the Great Depression suddenly arrived Simmons was caught entirely unprepared. Long on assets and short on credit, he found his business opportunities paralyzed. Nearly fifty years later, he related his predicament to his young grand-nephew, Ahmed Shadeed. He had helped Ahmed, a Vietnam veteran, secure a job with a Tulsa oil company, and was surprised to see the recently employed young man in the parking lot with a brand new Mercury Cougar. "You're driving a big car," he told Ahmed. "How much do you owe on it?" Ahmed admitted he had not even made his first payment. "Back when the Depression hit," Simmons said, "I had a brand new Pierce-Arrow. Then everything went bad. I was near starving. I couldn't eat that car, and I could barely sleep in it. A big car wasn't much good to me then. You try to sell that car." Ahmed did.[16]

Poverty was insufferable to a man like Simmons. Yet with the Depression his real estate and oil brokerage business completely dried up. Rural Oklahoma, as documented in John Steinbeck's *Grapes of Wrath,* was among the areas most cruelly hit by the economic crash. Banks foreclosed on farms by the thousand. Simmons had plenty of property to sell, but nobody had money to buy. Oil leases were even more impossible to peddle: nobody was willing to risk capital to drill new wells.

Simmons had to do something. He knew his parents' farm would always keep his family from hunger, but the last thing he wanted was handouts. J.J. III, the couple's oldest son, recalls his father's next move. "Oil had been struck in east Texas, and there were quite a few blacks down there that had oil on their property," he says. "I specifically remember my father giving my mother $100 and telling her that he did not know when he was coming back, but that he would not come back until he had turned some kind of deal in Texas. That gave me a terrible sense of insecurity. We lived in a very nice home on Court Street and the Depression had caught him with a 1929 Pierce-Arrow, which was probably the finest car in all of Muskogee, and which I as a child was very proud of. In those days practically everybody had accounts where you could charge groceries and pay them out by the month, which I presume is what we did. I don't think anybody realized that we were in that kind of predicament. Dad was probably gone a month, and I'm certain that was the lowest point in his life, and mine, too."[17]

In 1930 Jake Simmons risked more than his time traveling through east Texas in a flashy car. During this period the area was one of the most racially hostile in America. Simmons never traveled alone. For protection—and to share the driving—he often teamed up with Robinson Hickman, a part-Creek, part-black classmate from Tuskegee. Hickman doubled as a witness when documents needed signing. Having gone to Tuskegee he was as conscious of his appearance as Simmons; the pair moved through Texas in business suits befitting Jake's costly car, and they watched each other's backs.[18]

Before the oil hawks descended, land in east Texas had been almost worthless. Except for the northern border most of the region's hilly earth is red, claylike, and scorched. The area yields crops grudgingly and offers little shade or water. It was a grim landscape, written off as a waste of time by geologists of the big oil companies. Then, in October 1930, an old wildcatter named "Dad" Joiner proved the majors wrong by making a seven-thousand-barrel-a-day strike near the small town of Kilgore. Further exploration found the well to be part of a much larger oilfield spread across the region. With the ascendancy of the auto age the find attracted a greater oil rush than even the Spindletop or Glenn Pool discoveries. Brokers, middlemen, and investors like Simmons and Hickman raced into east Texas. Almost overnight the landscape became a forest of oil derricks packed so closely together they overlapped each other on hundreds of farms, town lots, and even in church yards.[19]

When they got to Texas Simmons and his partner found they were in no position to compete with the likes of H. L. Hunt, who had bought up Dad Joiner's well and much of the action in the area. Option prices for leases were high, and the two men had little cash. The weeks passed hopelessly. Simmons considered turning back, then remembered the vow he made before leaving Muskogee. As he traveled through the area talking to black landowners who had made considerable fortunes leasing their meager farmland for exploration, an idea dawned on him. Listening to constant tales of lynchings, beatings, disenfranchisement, and harassment, he realized that a lot of the

Texans would be happier living in Muskogee County. If he could convince them of this and sell them land in his part of the country, he, too, would be able to cash in on the east Texas oil boom.

Persuasion was never a problem for Simmons. Scarcely a month after leaving home he was back, with a handful of new friends. Selling farmland to oil-rich black Texans had not been his original intention, but he was a flexible businessman. "I was very surprised when he came back to sell people farmland," recalls J.J. III, "because that wasn't the idea at all."[20]

Jake Jr. did not let his new clients know this. To them he was an experienced farm broker. He introduced them to prominent members of the African-American community, showed them all the black professionals downtown, walked them through stores, and brought them to successful farms and ranches, like his father's. He also had plenty of former Texans to introduce them to: nearly one-quarter of all blacks living in Oklahoma had been born in Texas. There was a reason, Simmons explained, that so many black Texans moved to Oklahoma and so few Oklahomans moved to Texas. His tour included the rivers and streams that irrigated the rich farmland, the luxuriant grassy fields where Texans sent their cattle to graze, and the Muskogee County clerk's office, where some blacks were registered to vote. He rattled off crop yields, described the quality of the segregated schools, and hyped the excellent deals which cash-hungry sellers were offering. Many of the black Texans with oil money had scrounged out livings on tiny plots. In Oklahoma Simmons offered them larger, richer farms—opportune places to invest their money before it was all spent. For people with cash, like Simmons's Texas clients, the prices would never be lower.[21]

Simmons brokered a few sales from his initial batch of prospects. The commissions carried him through the first wave of the Depression. Over the next decade and a half, he returned to Texas again and again, both for oil deals and to find buyers for many other Oklahoma properties. Edythe Rambo Fields, a retired Muskogee schoolteacher who evasively describes her age as "seventy-something," grew up in Marshall, Texas. She recalls that her father, a farmer and Baptist minister, made a lot of money from oil that was found on his land. "In about 1936," she says, "Mr. Simmons came in and just talked him into moving. He said Oklahoma was the place to come. My father and mother came up here to look it over, and they bought some land he showed them at Boynton, Haskell, and here at Muskogee."[22]

Simmons had an uncanny nose for clients. Soon after Mrs. E. C. Ryan and her husband sold off the oil royalties on their farm in Kilgore, Texas, he appeared at their door. "He was a real estater," Ryan recalls. "He was a good salesman and influenced us. He said Muskogee County was a good place to live, that it has nice schools and nice people. We bought 240 acres. We decided to move out here and raised cotton and corn and had lots of cattle."

Simmons's services did not end with the commission check. In some ways he acted more like a colonizer than a broker. He kept in touch with many of the families he brought to Oklahoma for the rest of his life. Ryan depended on him to help manage her financial affairs. "I always felt free to go to him if I needed information concerning business problems, if there was something I didn't understand," she says. "We were always free to call on him for advice."

Many Texans with oil on their farms were far less fortunate than the Ryans and Fields. White brokers and oil operators often treated contracts with black landowners as meaningless. Sometimes well operators cheated on their production-level reports. More commonplace scams involved dishonest brokers, who pocketed royalties that landowners were legally entitled to. White hustlers acted as brazenly as they cared to, assured that the aggrieved party, if black, had about as much chance of winning a lawsuit against a white man in Texas as he had of being elected governor.[23]

Blacks were forbidden to serve on juries in east Texas. They never worked in law enforcement, were rarely attorneys, and were generally unable to vote. There was no challenging the system. The impossibility of legal restitution added to the natural distrust farmers had for smooth-talking oil brokers who bore handfuls of cash and conveniently prepared contracts. Some preferred just to keep farming their barren property rather than allow strangers the right to profit from it—and destroy the land's sparse productivity in the process. Taft Welch, now chairman of the Western National Bank of Tulsa, invested in the east Texas oil rush. Black landowners, he says, "didn't trust the white man, and damn well plenty reason not to. Because actually most of them ticket men [brokers] would steal; they wanted to screw up the leases and not give them a dollar. They'd represent one thing to get them to sign, then gyp them and take the main portion of their royalty. There was all kinds of skulduggery going on."[24]

Tales of dishonesty created many holdouts in black neighborhoods. Simmons realized that if he could persuade reluctant farmers that they would be given a fair deal, he could create a niche for himself in the east Texas oil play. On his next trip to Texas he did just that. His first step was to find an attorney in the white community willing to defend the legal claims of African-Americans ("title clearance" was always a major issue—black claims to valuable property were regularly contested). Cecil Storey was his man. At one time the oldest person in the state legislature, Storey was an ornery, tobacco-chewing, rough-talking redneck with a sense of honesty. He liked Simmons's determination and stood up for the black clients the Oklahoman brought to him, helping them win title battles with covetous white neighbors.[25]

Representing blacks in Texas was a risky business. John Shillady, the NAACP's white executive secretary, found that out in 1919 when he was sent to the state capitol to argue that the legislature had no right to close his organization's Austin branch. On his way from a meeting with belligerent local politicians he was attacked by a vicious mob and beaten unconscious on the steps of the courthouse. Langston Hughes, in *Fight for Freedom,* his 1962 history of the NAACP, wrote that Texas governor William P. Hobby responded to an NAACP appeal to punish the assailants with a telegram stating that Shillady "had gotten exactly the kind of treatment he deserved and that the same treatment awaited any other white man who dared interfere with the way Texas controlled its Negroes."[26]

Little had changed when Simmons began brokering black people's land near Kilgore and Longview. "East Texas was a place where black people had no rights," recalls Donald Simmons. "A black person could be killed and essentially there would be nothing done about it. I've been through towns in Texas where they burned down the whole courthouse so they could kill a black man in it. Just burned the whole goddamned thing down. Yet he bought a lot of leases in east Texas when it was flat dangerous for a black man to be down there. It took an enormous amount of courage and an enormous amount of confidence."[27]

The threatening racial environment was made even more dangerous by Simmons's quick temper and his refusal to be treated disrespectfully. As a boy of twelve, J.J. III remembers accompanying his father into an east Texas drugstore to buy a newspaper. "What do you want, boy?" the young clerk behind the counter snarled. "My name is Jake Simmons," he said. "I don't care what your name is," the clerk demanded, "Either you tell me what you want or you get the hell out of here." According to J.J. III, "Dad reached behind the counter, picked him up and slapped him a couple of times and we both walked out. Which was most unusual in those days in east Texas."

To make money in Texas Simmons needed more than an attorney and the guts to hustle leases. He needed to find a drilling company which would honor its lease commitments to both him and the black landowners he received contracts from. He approached a prominent landowner named B. F. Phillips, who controlled three thousand acres in Longview and a large country store where many blacks traded. Phillips, one of the biggest players in the lucrative Hawkins oilfield, was shocked when Simmons offered to broker deals for him. He had never dealt with a black broker before. "Boy," he said, "you don't know anything about the oil business." Jake said that he could get the Texan the best leases in the area, and added that he resented being called a boy. Phillips invited him home that evening, where the two men discussed business sitting on the porch. Phillips quickly developed a sense of admiration for the straightforward young broker. "I didn't know that you objected to being called 'boy,' " he said, "because we call most of your people 'nigger' or 'boy.' " "Well," Simmons explained, "I'm not one, and I resent being called it."[28]

Terminology was more than semantics to Simmons. It represented the level on which one person viewed the other, and the man who acknowledged the lower position could rarely avoid internalizing this distinction. "You always remember," he would tell his children, "that the color of your skin should not make any difference. You are equal to anyone, but if you think you're not, you're not."[29]

Phillips agreed not to call Simmons "boy." He also agreed to buy whatever leases Jake could get him in a designated area of the Hawkins field. Through the next decade Simmons sold many of his Texas leases to Phillips. As his reputation grew, black landowners were increasingly likely to welcome the oil industry's first important black lease broker into their home. Most deals were concluded around the kitchen table. Although Robinson Hickman or his brothers Jim or John sometimes joined him, Jake did most of the talking. His assurance that lease signers would receive their one-eighth royalty carried considerably more weight than the word of the average white broker. Landowners also received cash on the signing: from one dollar per leased acre for unproven land to $100 an acre for land near productive wells. Additional performance payments were also spelled out in the contract: roughly $300 in "damages" was promised for each well eventually drilled, plus another $100 for use of an area to build an access road. The farmers generally received a minimum of a few hundred dollars at the signings, money which went a very long way during the Depression. If commercial quantities of oil were found, they received steady incomes. Many were also hooked up to all the free natural gas they could use. Simmons's profits flowed from the small percentage interest he sometimes retained, but more often he made money from the outright sale of a lease, or assemblage of leases. On a good deal he could flip his lease to an oil company for two or three times what he paid.[30]

It was important to Simmons to be associated with a backer like Phillips. Jake was a very conservative investor, and generally avoided putting his own money on the line. Taft Welch, who knew of Simmons's reputation, believes that he typically ended up with a 5-percent override. Welch thinks that because he never demanded more than this "very reasonable" percentage, Jake became more popular than other brokers who, when they had a good lease, would try to "skin a deal" by demanding a 25-percent royalty. Welch says that oil companies allowed Simmons his 5 percent and then reimbursed his out-of-pocket investment. Although not entirely accurate, Welch comes close to the mark when he says, "He never lost anything but his time. He never used his own money, or tried to buy into anyone else's deal. He wouldn't invest in a twenty-dollar gold piece. But he'd take your money, and give you some kind of a split on it."[31]

Jake's friendship with Phillips helped him secure leases from a few white landowners as well as blacks. It also helped protect him from those who were enraged by the sight of a well-dressed black businessman making out in their home territory. Phillips was one of the major oil multimillionaires in the region, and his support went a long way in east Texas. "He was like a godfather to my dad," recalls Donald Simmons. "In any community everyone knows who the big man is, and I'm sure that the reason my dad did not get killed was because everyone in the white community there knew that he was a friend of B. F. Phillips."

Phillips was one of many important whites whom Simmons befriended. Before he became involved with Phillips in the Hawkins field, he brokered leases to numerous drillers in the "east Texas field," which became a hot spot about 1933. In this region he found local support from a prominent attorney named Sam B. Hall, Sr. Hall was a wealthy oil investor and part of an influential law firm in Marshall, Texas (not far from Kilgore and Longview). Jake was having one of the firm's lawyers look over a lease for him when he met Hall, a politically connected attorney who later became county attorney and district judge. The two men hit it off well and became close friends. Sam B. Hall, Jr., who served five terms as a Democratic congressman from the first congressional district of Texas (he was considered one of the wealthiest members of Congress) is now a federal district court judge in east Texas. He has a clear and affectionate recollection of Simmons from the Oklahoman's frequent visits when he was a boy of ten. "Simmons worked out of my father's office whenever he was in the area," Hall says. "He would go to Longview and make trades, then come back to the office and my dad or his secretary would help him prepare leases. When he was in town he'd come to our home for lunch or dinner. I admired him a lot. He was the ultimate gentleman, always dressed to perfection with a coat and tie. He always wanted me to go back to Oklahoma with him to visit and I really wanted to, but my parents said I couldn't go because I was too small."

Congressman Hall admits that it was "highly unusual" for a black man to be dining in the home of a prominent white family at that time, but he explains, "I don't think it was ever thought of one way or another. They were just close friends. He exuded honesty and integrity, and that went a long way in his ability to infiltrate with the people of east Texas at that point in our history."[32]

Simmons made a similar impression on oilmen in Oklahoma. He peddled leases to some of the biggest in the business—industry legends like Harry Sinclair, William Skelly, and Frank Phillips. Giant oil companies like Texaco, Sun Ray DX, and Phillips Petroleum also did business with him. Sometimes large operators fronted him money to buy up blocks of leases in a particular area which they wanted to explore. More frequently he was a free agent. Hustling was the name of the game during the Depression, and Simmons hustled hard. He moved throughout Oklahoma, Arkansas, Louisiana, and Kansas, as well as Texas, in search of properties and buyers. Through it all he made sure to keep his reputation clean. A person's word in the oil industry was his stock-in-trade. "My dad never signed a contract with any of these men that he did business with," explains Donald Simmons. "And they would trust him with unlimited sums of money on his word. You have to look at that to say there's a whole lot in this country that's right even when it's at its worst."[33]

Donald believes that many of the oilmen his father did business with would have disagreed angrily with their black associate if they had ever discussed civil rights with him. But Jake segregated his political and business activities. He influenced conservative oilmen not by talking to them about oppression but by conducting himself in a manner which won respect for both himself and his race. "The northern white liberals have a great deal of trouble understanding how people did things in the segregated South," Donald says. "The so-called right-wing conservative businessmen didn't have any problem understanding that anybody who's worth anything would want to be in control of his own destiny, would want to make the same thing that worked for most of America work for him. There were always sympathetic people willing to give a man a chance, and that's one of the great things about this country. If there hadn't been white people who were sympathetic and whose hearts were in the right place, I'm sure all the black people in this country would be dead."[34]

Jake Simmons did more than accept an opportunity presented to him by others. He generated opportunities, achieving what his mentor, Booker T. Washington, had urged black Americans to do. "We must create positions for ourselves," Washington said, "positions which no man can give us or take from us."[35]

5

HOLDING OUT FOR DIGNITY

If you go out to the Simmons place I'd suggest you get in touch with your next of kin first, 'cause that's one black man who would kill you.[1]
—Archie Wright, a white friend of Jake Simmons's, speaking to a whipping party forming to "teach" him a lesson

In any system crumbs are allowed to fall. No matter how far those at the top are from those at the bottom, they are always sure to allow a few scraps to filter down. Although Oklahoma virtually shut African-American citizens out of the political system during the fifty years that followed statehood, the power structure allowed an occasional black voice to be heard—provided it said the right thing.

Although a few counties, like Muskogee, tolerated black voters, most of Oklahoma's political crumbs took the form of patronage jobs and promises to upgrade the invariably horrid conditions at state-run black institutions. Jake Simmons's first major confrontation with the system was over the fate of such an institution.

In 1936 Oklahoma governor E. W. Marland ran for the U.S. Senate with a promise to black supporters that he would improve the scandalous management of the state's most important black facilities. Three large institutions—a detention center for juveniles, a home for the mentally ill and blind, and an orphanage—were all located in Taft, an important black town ten miles from Muskogee.

The stewardship of Taft was Oklahoma's choicest black patronage plum. It was held by a local celebrity known as "Major" H. C. McCormick. In 1932 McCormick was an obscure assistant cashier at the state's only black-owned bank in Boley. Three armed bandits, including a partner of the notorious robber "Pretty Boy" Floyd, took over the bank, locked McCormick in a vault, then shot the bank president dead. While the bandits were busy emptying the tills McCormick opened the vault from inside and jumped out with a high-powered rifle hidden there. He fired on the gunmen and managed to kill all three of them. McCormick parlayed his ensuing fame among Oklahoma blacks into a political patronage career, capped by a 1935 gubernatorial appointment to run one of Taft's facilities. He soon managed to wrest control of all three Taft operations. Solidifying his unbridled power as "business manager," he built a lucrative fiefdom and doled out jobs to relatives and sycophants. White political benefactors gave him free rein as long as he came up with hundreds of votes at election time.[2]

Rumors of exploitation and scandal at Taft slowly reached leaders in the state's black community, who lobbied the governor for an investigation. Governor Marland liked McCormick, but because of his campaign promise he had to appease other blacks by giving the impression he would clean things up at Taft. He appointed seven of the state's most prominent African-American leaders to "an unofficial board of regents" to advise him. Included on the board were Jake Simmons, Jr.—listed as "president of Simmons Royalty Company"; Roscoe Dunjee, crusading editor of the Southwest's most important black newspaper, the *Black Dispatch;* and J. W. Sanford, president of Langston University.[3]

Although the board was granted investigatory powers, little was expected from their probe. Any recommendation made by Taft's new regents had to be approved by the state's all-white "Board of Affairs," which governed Oklahoma's institutions and enjoyed a very cozy relationship with Major McCormick. "Responsible" black leaders were expected to recognize this, and play out their parts without making waves. It was unimaginable that a prominent black could also be a rebel. The system recognized two kinds of Negroes: insiders and outsiders. The patronage game allowed captains, like McCormick, lieutenants, and foot soldiers. Governor Marland expected the board members to act like aspiring lieutenants and carry out their duty to the man who appointed them.[4]

But Simmons and Dunjee did not play by the old rules. They were part of a new generation of African-American leaders whose objective was progress, not accommodation. Their private businesses brought them a measure of independence unknown to the average black worker. Without the constricting economic pressure of white employers the two were free to gravitate to the top of the region's civil rights movement. Both were influential members of the state's NAACP (Simmons joined soon after returning from Tuskegee), and neither was the sort to turn from injustice.

At first the board's relationship with Major McCormick was cordial. As they started their investigation in September 1937, he served them a tasty Sunday dinner and promised that all Taft records would be made available within the next few days. The committee, acting on tips from insiders, gave the lord of Taft a list of employees and "inmates" it wanted to question, and arranged for a tour of the three institutions.[5]

Simmons and the other board members found the children at the mental institution and orphanage without shoes or clean underwear. During months of private hearings they also discovered that Taft's officials were engaged in something resembling a slave trade. Black girls were regularly "farmed out" to work long hours as live-in housekeepers and nannies for white people—either without pay, or for two or three dollars a week. Some were occasionally even taken out of the school to be "adopted" by white men—one from as far off as Indiana. The board was never able to determine whether this was done to line the pockets of Taft officials or reduce overhead at the institutions: the regents were denied access to McCormick's financial records.[6]

They were, however, able to question residents at the institutions. Seventeen-year-old Eva Jefferson, one of two pregnant orphans, testified that she had been raped by a white man who had been entrusted with her "care." She said that she worked for her temporary "father" from early in the morning until 11:00 P.M., and was promised two dollars a week, but paid nothing. Simmons, a member of the questioning panel, was shocked by her account.[7]

Equally alarming was testimony by Louella Bruner, nurse at the girls' orphanage. Although boys and girls were kept apart at the institutions, Bruner told the committee that ninety-one of her girls suffered from syphilis.[8]

By the time Simmons got Major McCormick to the hearings he was furious. He demanded to know if the Taft superintendent remembered being notified that Eva Jefferson had been raped by a white man. McCormick stonewalled. "I don't remember," he said.[9]

The board was beginning to make McCormick uncomfortable. They argued that control over the three institutions needed to be decentralized, as it had been before McCormick took over. Three qualified superintendents, they noted, would do the job far better than one inexperienced headman. They redefined the Major's job title, calling him "director of interlocking departments" instead of superintendent.

McCormick and his cronies fought back. Using their far smoother political connections, they rallied support from Taft's official governing body, the state Board of Affairs. Employees of the three institutions appeared before religious and secular groups across the state requesting that letters be sent to the governor opposing the board of regents. Governor Marland met personally with Taft officials in Oklahoma City to hear their "grievances" against the very investigating body he himself had appointed.

On November 3, Roscoe Dunjee met privately with Governor Marland and Mable Bassett, commissioner of charities and corrections, in the "Blue Room" of the Oklahoma City state capitol. He tried to convince the two that his unofficial board needed legal authorization to effect much-needed changes at Taft. He pointed out that the white board of regents at Langston, the black university, had unchecked fiscal authority there.

Marland balked. The Taft board consisted of blacks. He had appointed them as window dressing, not to exercise power. Dunjee's request was flatly denied.

The next day Dunjee resigned from Taft's board of regents. In the next edition of the *Black Dispatch* he blew the whistle on the governor's partronage mill. He headlined a front-page article" 'DUMMY BOARD' SAYS DUNJEE IN RESIGNATION FROM REGENTS AT THREE TAFT INSTITUTIONS." He pulled himself off the board but not out of the picture. The next week, controversial transcripts from the board's hearings started making their way into the *Black Dispatch,* with screaming page-one headlines like one which demanded to know, "WAS EVA MAE JEFFERSON RAPED?"[10]

Simmons, slightly younger and less politically experienced than Dunjee, stayed on the board a few months more. He naïvely believed that once the regents' investigation was over the state's Board of Affairs would dismantle Taft's corrupt hierarchy and improve conditions for hundreds of destitute blacks there. Once he brought the truth to light, he reasoned, change would follow.

Not until January 1938 did he realize he was wrong. At a board of regents hearing over an inmate's rape Simmons got into an argument with Mable Bassett, the powerful corrections chief. Bassett launched into a tirade against Taft's outspoken critic because Simmons neglected to stand when she entered the room. Then, during the hearing, Bassett broke off into repeated private conversations with Major McCormick, her close friend. Simmons objected, saying that the other members of the board should be privy to their discussions. Bassett demanded an apology. Simmons refused. Her partiality toward McCormick had become all too evident. It was finally clear to Simmons that it was futile to expect officials like Bassett to act on the mountain of evidence against the corrupt superintendent. He angrily resigned from the "dummy" board of regents, and printed an outraged letter in the *Dispatch* charging that "this board is wholly without administrative authority to effect a constructive program for either of said institutions."[11]

Public opposition to the state's mismanagement of Taft grew. A week after Simmons's resignation, fourteen hundred delegates at the Oklahoma convention of the Veterans of Industry of America (an integrated labor organization) voted overwhelmingly to "denounce and condemn the horrible conditions exposed" at Taft. The resolution requested that Governor Marland "give the Negro board of regents a free hand in managing these three institutions, and end the intolerable situation by which the state board of affairs is making them political footballs and patronage hangouts."[12]

The negative publicity was mounting. Instead of wielding a rubber stamp the Taft board of regents had documented horrible abuses and insisted on real changes. Governor Marland decided to end the charade before it made his administration look even worse.

In early February, Taft's regents received a letter from the state Board of Affairs, which had appointed McCormick to begin with. The letter did not ask for recommendations; it dictated exactly who were to lose their jobs for the scandals and who were to take their places. McCormick and his top cronies were left untouched, while the full-time board of regents secretary, Dr. A. P. Bethel, was to be fired.

Another board of regents member resigned immediately in protest. Governor Marland was now able to pack the board with enough cooperative new members to form a majority. On February 15, the newly realigned board of regents met for the first time. By a vote of four to three they agreed to fire those employees the state wanted them to fire and recommended the return of Major McCormick to control all three of Taft's institutions.[13]

Marland's administration managed to cover up an embarrassing situation, but it was a close call. Simmons and Dunjee were branded permanent outsiders—independent thinkers unwilling to remember their place. It would be three decades before Simmons was appointed to another state government–appointed committee.

Simmons again came into conflict with the power structure only a few months later. His opponents this time were prominent Muskogee whites, and the battleground was not a segregated institution but the very system of segregation itself.

On September 15, 1938, the city of Muskogee held a public referendum to win legal approval of a $275,000 bond issue. All the money was to be spent improving white schools and building a new white junior high school. The federal Public Works Administration, a New Deal agency, promised the Muskogee Board of Education $225,000 in matching funds for the school program. The city's leaders pushed hard for the referendum, which passed by a two-to-one margin.[14]

Jake and Eva Simmons objected to having to pay additional taxes to finance schools their three children would never be allowed to attend. They discussed the situation with Charles Chandler, a close friend of Simmons's and Oklahoma's most important civil rights lawyer. Simmons had been helping Chandler finance a landmark Supreme Court challenge to Wagoner County's grandfather clause; he even brought his young sons with him as he went door-to-door collecting nickels and dimes from black Muskogeeans to help pay court costs. (The case, *Lane* v. *Wilson,* struck down discriminatory voting practices in counties across the country.)

Chandler, who was backed by the NAACP in the Lane case, was similar to Walter White, one of the earliest and most effective leaders of the organization, in that both men had skin so light in color that they were frequently mistaken for Caucasians. Chandler grew up in Cleveland, received a degree from Yale Law School in 1923, then moved to Muskogee. His office was in the same building as Simmons's, on South Second Street, in the heart of Muskogee's black business district. Like Simmons, Chandler distinguished himself early on as both a brilliant professional and a man with a fierce sense of racial pride. (Once, when a man called him "nigger" in a courtroom, Chandler broke a chair over his head.)[15]

Chandler and his wife, Elsa, were among Jake and Eva's closest friends. The two men were expert bridge players, and the couples met regularly for games. "Dad's bridge playing spilled into his life," recalls J. J. Simmons III. "It was never just making the bid. To get a slam was what he went after all the time. He would go down in a minute. It was phenomenal how many times he did, because he had that confidence."[16]

Simmons was no less courageous when it came to taking on the system. With Chandler's help he and Eva prepared a scathing legal argument against the school bonds. The day before the referendum, they filed a lawsuit in Eva's name in federal court, seeking to bar the sale of the bond issue on the grounds that it was unconstitutional.

"In the entire proposed outlay of a half million dollars of public funds," the lawsuit charged, "no provisions have been made for the expenditure of so much as one dollar for the direct or indirect educational benefits of any Negro child, although it is proposed for Negroes to pay taxes to discharge such bonds."[17]

The lawsuit named as defendants the state of Oklahoma's district attorney, as well as Muskogee's mayor and its Board of Education. In addition to challenging the bond issue the suit charged that the school district had no right to call a referendum, since the very state laws which created a segregated school district discriminated against blacks, and were therefore unconstitutional.[18]

Had an antisegregation suit been filed at any other time it might have gone unnoticed. But Simmons, through his recent experience at

Taft, had gained a better understanding of opposition politics. Black "insiders" got nowhere asking local officials for help; the only challenges that meant anything were those that were timed right and came from the outside—meaning the federal judiciary. By tying his lawsuit to a sensitive bond issue he was able to hit Muskogee's power brokers where it hurt them most: in their pocketbook.

Because of the Simmons lawsuit the placement of Muskogee's school bonds had to be put on hold. The area's conservative banks were unwilling to risk $275,000 on bonds which they might never be able to issue. Moreover, the $225,000 in federal PWA funds were contingent upon the bond money's availability, and the agency had set an October 31 deadline for eligibility. If the case dragged on too long, Muskogee's school board worried, the federal funds could be lost.[19]

Because of the constitutional issues involved, Muskogee's federal district judge Eugene Rice decided that *Simmons* v. *Muskogee Board of Education* had to be heard by a three-judge circuit court. Two other federal judges from Oklahoma were brought to Muskogee to hear the case.[20]

On October 1, two days before the trial, the *Muskogee Phoenix* ran a front-page article informing readers what was at stake. The headline read, "State's Separate School Law Faces Ruling Monday." The article said that the suit had been filed by the NAACP, in an action taken by "Willie Eva Simmons, Muskogee Negro woman." It continued: "Charles Chandler, Muskogee Negro attorney retained by the association [NAACP] to prosecute such suits, announced that its purpose involves much more than 'merely throwing out the bond issue.' It is designed, he said, to invalidate the Oklahoma public school law and put Negroes and whites in the same classroom."[21]

Although the NAACP was not actually behind the lawsuit the *Phoenix* was eager to make the case seem like part of a larger conspiracy. Attorneys from Oklahoma attorney general Mac Q. Williamson's office warned *Phoenix* readers what was at stake: "Williamson's assistants frankly admit that if the case holds good the school laws in all southern states would be voided. Such laws already have been upheld by state courts, but have never been taken to federal courts."[22]

That Simmons and Chandler were among the first ever to challenge school segregation in federal courts branded them as radical troublemakers. On the day of the hearing the *Phoenix* ran another front-page article stressing the severity of the threat posed by the rabble-rousers. With the subhead "Suit in Name of Muskogee Negro Woman Challenges Entire System; Appeal Likely," the paper again accused "national Negro organizations" of financing the lawsuit. According to the *Phoenix*, "J. Harry Johnson, assistant state attorney general, said a victory for the Negroes would have the effect of permitting Negro and white children to go to school together. 'If they win, it would break down the separate school system throughout the south,' Johnson said. 'We contend that our system has been recognized and upheld by practically all state courts, including the state supreme court.' "

On October 3, 1938, Muskogee's graceful federal courthouse, occupying a square block on North Fifth Street, was filled with a small army of government officials and their attorneys. The elegant second-floor courtroom, with its thirty-foot ceilings, shiny brass fixtures, richly upholstered maroon leather seats, and sturdy oak doors, conveyed a sense of sanctity to the presiding judges. They represented the federal judiciary, the highest arbiter in the land, and if the black plaintiffs questioned the objectivity of the white man's court, they had only to look upward for solace. There, carved in stone for all time, were the words of Daniel Webster: "Justice is the great interest of man on earth."[23]

The attorneys for the school system presented reams of statistics and a procession of experts, all supporting their contention that the black schools of Muskogee were in tiptop shape. The area's 1,916 Negro students, they said, represented 23 percent of the total in the school district. In the 1937–38 fiscal year the average expenditure for black high school students nearly equaled that of whites. In elementary schools there was more of a disparity: $60.21 was spent per white pupil, and $38.79 per black. The average white schoolteacher earned $1,172 annually; the black teachers earned $887. (These comparisons were later sharply disputed by Chandler, who claimed much wider discrepancies.) A state inspector testified that the black schools were in "acceptable shape," and the superintendent of Muskogee schools told the judges that the black high school was less crowded than the white one.[24]

When Eva Simmons took the witness stand she told a different story. The black schools, she said, were in "very poor" condition; they were badly heated, had obsolete toilets, and no landscaping or playground facilities. The white schools, she pointed out, had gymnasiums, while the black schools used their auditoriums as assembly halls, gyms, and libraries. Her argument that part of the $275,000 earmarked for white schools was needed by black schools was supported by a different set of statistics. Although black students made up 23 percent of the school district, the value of their school buildings and improvements was 13 percent of the district's total value. The value of their school equipment was just 9 percent of the total. This meant that the county had previously allocated roughly half the amount per black student for capital expenditures that it had for whites. If the new funds were to be spent as announced, she argued, this disparity would widen even further.[25]

Attorney Charles Chandler presented similar testimony from a local black teacher and a school official, then concluded that although blacks were taxed as much as whites, the money spent for their children's separate education was far from equal. Therefore, he argued, it was a clear violation of the Constitution's Fourteenth Amendment, which extended American citizenship to blacks and forbade any state from denying them equal treatment under the law.

After Chandler presented his case government officials filed several motions to dismiss the lawsuit. In overruling the motions circuit judge Robert Williams complained that $97,000 of the federal grant money was to be spent building a sports stadium for white students. "You build a costly stadium while these Negroes are crowded like they are. I don't believe it is a credit to the white race. . . . When you were getting that PWA grant, why didn't you try to get some money for these Negroes out here?"

Although the court refused to dismiss the case it agreed to a conditional go-ahead for a school bond auction, scheduled for that very evening. They could have the auction, the judges ruled, but the state attorney general was explicitly restrained from approving the bond issue.[26]

The restriction cast a grim pallor over the auction. All but three of the many bonding companies and banks in attendance withdrew from the auction when they heard about the complications caused by the lawsuit. The Citizens National Bank of Muskogee bought the bonds for what the *Phoenix* called "an unusually low rate of 2.87 percent."[27]

Jake Simmons was becoming very unpopular in Muskogee's corridors of power. City and state attorneys assured the politicians of the likelihood of a victory in Muskogee's federal court, but warned of long-term problems if Simmons appealed to the Supreme Court. Such an appeal would tie things up well beyond the Public Works Administration's October 31 deadline for matching funds, and Muskogee officials would be forced to beg for one extension after another. An appeal would also postpone the delivery of the bond money and potentially even result in an outright defeat for the city. The stakes were high. Yet if Simmons could be persuaded to drop his appeal, all these problems would vanish.

A group of civic leaders decided it was time to do a little arm-twisting. They checked the city records and found that Simmons, like most Muskogee residents, owed a considerable sum in paving taxes for the comfortable home on Court Street which he owned. That year Muskogee had undergone the most extensive paving program in its history. Taxes were increased and each homeowner was required to pay a share of the costs over a period of years. Simmons's bill for paving taxes amounted to $2,500.

According to Donald Simmons, his father was called to a meeting of prominent bankers and politicians at one of Muskogee's largest banks. "We agree with you, Jake," one of the men told him. "This is a terrible system when a man of your ability can't go to places that he wants to, and you can't send your child to the school of your choice. So we think it would be better if you would move up North, where your family would enjoy all of those benefits." The remark was more a threat than a suggestion. The paving taxes Simmons owed were being called in—he had just a few days to come up with the $2,500, or his house would be foreclosed. There was, of course, the lawsuit. If Simmons and his wife agreed to drop it, the bank would extend them credit for the taxes. If not, they warned, he could be sure that no bank in the state would lend him a cent.

Simmons asked for a day to think it over. He asked Eva what she thought. "We can always build a new home," she said, "but when it comes to what's right and wrong, we can't compromise."[28]

The Simmonses never had to move. Not everyone in the white community wanted to drive them out of town. When word of the banker's ultimatum reached Ernest Anthis, a white business associate, he phoned the black oilman. "Quite frankly, Jake," Anthis said, "I don't think your fight is worth it. But I heard what those no-good SOBs tried to do to you. I'll lend you any amount of money you need to pay off those taxes."[29]

The next day Simmons met the city leaders at the bank. "I'll move," he told them, "if you can turn the sun into the moon." And he walked out.[30]

It was more than bold words which allowed Jake Simmons to survive such confrontations. Racial supremacists and Klansmen viewed civil rights agitation with as much antipathy as Communist Party organizing. Not all of Simmons's detractors confined themselves to financial pressure in trying to run him out of town. On several occasions his windows were blown out in the middle of the night by shotgun blasts. Then there were the phone calls at all hours, in which mysterious, hateful voices threatened, "We're gonna get you, nigger—you're a dead man! Get outta town by Sunday or we're gonna blow up your house, with everyone in it."

Simmons did not waver. He kept a .45 automatic pistol within reaching distance anytime he was home. A powerful rifle sat in the closet for sharpshooting. He taught his wife to shoot, and before his sons were teenagers, they learned, too.

Word got around. Archie Wright, another white friend of Simmons's, was sitting in a pool hall when he overheard a group of rowdy men announce that they were "gonna teach this nigger Simmons a lesson." Wright warned them, "If you go out to the Simmons place, I'd suggest you get in touch with your next-of-kin first, 'cause that's one black man who would kill you."

The mob found this hard to believe. "You mean to tell me he wouldn't be scared if a whole group of us came to his door?" one of them asked.

"Not only that," Wright said. "He's got a thirty-thirty [rifle] out there; he'll kill you before you get out of your car."

Wright called Simmons to warn him but it was hardly necessary. The men never came.[31]

At five feet eleven inches and 190 pounds, Jake Simmons was feared by men far larger than himself. Most of his close friends and relatives recall his hair-trigger temper; they abound with tales of him smacking someone who slapped him, called him "nigger," or was discourteous to his wife.

Donald Simmons believes that the lesson of the Tulsa race riot was not lost to his father—or his father's enemies. "He was driven by a mission and didn't think about death," Donald explains. "His theory was he would defend himself and sell his life very dearly. I think that's part of the reason he didn't get killed. When you internalize that you are willing to die for what you stand for, then the person that wants to try to take something from you is going to have to give everything he has. And cowards don't operate in that fashion very well."[32]

Simmons was always conscious of the example he set for his family and community. "My father didn't compromise," Donald Simmons says. "When he pushed that lawsuit, he scared the death out of some of the black people here because he just didn't know how to back down. If he had to lay it on the line, he would. That's the way he taught us: if anything's worth believing in, it's worth fighting for. That's an old American tradition."[33]

But Jake and Eva Simmons were up against another old American tradition: racism. On October 12, the federal court in Muskogee denied Simmons's application to forbid the sale of the bonds and upheld the constitutionality of the separate school law. The decision was big news in Muskogee. Bigger news, however, was that Chandler would be appealing the case to the Supreme Court. The front page of the October 13 *Phoenix* was headlined APPEAL FROM SCHOOL BOND ISSUE IS HINTED.

The article's subhead noted that "Delivery of Securities May Be Help Up Indefinitely; Discrimination Unproven." The *Phoenix* complained that the NAACP "instigated the appeal" even though the judge had clearly ruled that the separate-school law did not violate any federally guaranteed constitutional right. The paper failed to mention that the judges also concluded that the district's black schools were in desperate need of funding. "There should be some improvement apparently," said the court's finding of fact, "as to the sewerage and toilets, adequate remedy by law being available for such relief." The judges believed that Eva Simmons's legal remedy should have been to petition the school board and county to improve the district's black schools.[34]

Simmons and Chandler did not agree. The fact that the black schools were obviously in need of improvements yet were slated to receive nothing of $500,000 in local school improvement funding was clear evidence that there was nothing equal—or constitutional—about "separate but equal."

The national leaders of the NAACP also didn't agree with the court's ruling. At the organization's New York headquarters, NAACP executive secretary Walter White and legal staffer Thurgood Marshall (who later became America's first black Supreme Court Justice) studied Chandler's complaint, then approved funding to take the case to the Supreme Court.[35]

On Saturday, October 22, 1938, the executive committee of the Oklahoma Conference of NAACP branches gathered for a three-hour meeting at the office of the *Black Dispatch*. Publisher Roscoe Dunjee, head of the state conference and for many years the Southwest's representative on the NAACP national board, reported in the *Dispatch* that the group listened to Simmons tell of the "many methods of intimidation practiced by Muskogee authorities in an effort to force dismissal of the suit."[36]

Simmons's lawsuit dragged on for half a year, paralyzing the city's plans for its bond issue. In mid-December, the Citizens National Bank wrote the Muskogee Board of Education that the bank's Chicago attorneys "take the position they cannot approve the issue until the litigation, which is now pending, has ultimately been approved of." The bank officers offered Muskogee back the bonds (which it had acquired but not yet paid) if the city could find a new buyer. If not, Citizens National would be pleased to pay for the bonds—once it received a "nonlitigation certificate." The board didn't even try to resell the bonds. The arrival of the new year found school officials and attorneys pleading with the Public Works Administration for a second extension of matching federal funds.

Chandler's Supreme Court appeal cited a number of errors in the district court's ruling. He argued that the judges admitted into evidence false statistics showing comparable expenditures for black and white schools. He also noted that the district judges themselves had acknowledged that the bond issue was a violation of the Fourteenth Amendment, since they had stated in their findings that both the black and white schools were in equal need of repair—while all of the $500,000 was going to the white schools.[37]

In retrospect it might seem that *Simmons* v. *Board of Education,* fifteen years ahead of its time, was sheer folly. But Jake Simmons's optimism permeated every aspect of his life. Odds never determined the way he chose his fights. Principle—and effectiveness—did. By holding up an important construction project Simmons and Chandler were engaging in a rare practice for black citizens of their day: they were hurting the system with its own laws.

Between *Simmons* v. *Board of Education* and Chandler's grandfather clause case, Oklahoma became a critical civil rights battleground in 1939. The decision to overturn Wagoner County's discriminatory voting practices was clearly in line with earlier Supreme Court rulings. The Simmons school segregation case, however, tread upon the holy ground of a state's power to keep the races apart. The Supreme Court had affirmed that this power was the domain of individual states in its 1896 *Plessy* v. *Ferguson* decision. To challenge and thus desecrate the "purity" of the Caucasian race by sending white children to integrated schools was far more than the nation's highest judges were prepared to do in 1939. Even a decade and a half later, after the court's historic *Brown* v. *Board of Education* decision ordered public school desegregation, an army of national guardsmen had to be dispatched to places like Arkansas to quash the rioting.[38]

On March 16, 1939, the district court received a decision from Chief Justice of the Supreme Court Charles Evans Hughes. The mandate affirmed the district court's decision, dismissing the appeal. Its foundation, the court noted, was "so unsubstantial as not to need further argument."[39]

The day after the ruling the school board money was released and the city prepared to buy a site for the white sports stadium. The system had demonstrated that "separate but equal" did not mean that blacks had any constitutional right to true equality.

But it also demonstrated something else: although Simmons had never been invited into the state's power structure he was there anyway, a force to be reckoned with in eastern Oklahoma.

6

THE BUILDING BLOCKS OF PRIDE

I'm not raising bootblacks—I'm raising presidents.
And the only shoes my boys are gonna shine
are mine and their own.[1]
—*Jake Simmons, Jr.*

There's a story about Jake Simmons, Jr., which those closest to him love to tell.

Donald, Jake's youngest son, was fast asleep one Sunday morning after partying away the previous night. When his father told the teenager to wake up the boy refused. "I'm not going to church today. I am just too sleepy."

"Oh, yeah, you're getting up," Simmons said. "How in hell can a black man stay in bed in the morning when white men rule the world?"[2]

Donald got up. With a father like his there was little choice.

Simmons made a clear distinction between disobedience and self-expression. He taught his three sons to speak up when they disagreed with anyone—including him. The family home was a battleground of debate, since he seldom agreed with their dissension. But lying was strictly forbidden. This was Eva's rule, and she repeated it again and again, warning them, "If you lie, you're a coward." To their mother, the boys knew, nothing was worse than a coward. "She was a stickler for the truth," recalls J. J. Simmons III, now in his sixties and a commissioner of the Interstate Commerce Commission (one of the highest-ranking positions held by a black appointee in Washington). "And boy, she got that into me. She had as profound an impression on my life as my father. Because that has an effect on me today; I still try to avoid telling a lie, even a small one."

Although Jake Simmons had a notoriously violent temper (he once beat up a racist milkman for refusing to address his wife as "Mrs. Simmons"), he never struck his own children. He was impressed by the ability of the Japanese to raise well-disciplined children without hitting them, and did his best to emulate them.[3] Eva administered occasional spankings, while Jake followed the example of his own father's "tongue-lashings." J.J., the oldest child, remembers, "He was the one who talked to you. There was no termination on his lectures. He would sit you down, and it was logic, the whole time. He loved to talk. I think many times a spanking would have been preferable to the lectures that you got."

Much was required of Simmons's children. During World War II, when the 45th Division's nearby training camp created a big demand for polished shoes, a white business associate asked him why he did not send his boys out to shine shoes.

"Why aren't your boys out there?" Simmons replied. His friend did not answer. "You have a view that any black child ought to be shining shoes," he said. "I'm gonna tell you what: I'm not raising bootblacks—I'm raising presidents. And the only shoes my boys are gonna shine are mine and their own."[4]

Simmons strived to develop in his sons the building blocks of pride within a racist society: self-worth and confidence. He allowed J.J. III to drive his car when the overgrown boy was ten. By the time he was thirteen, in 1938, his father called from east Texas to have him deliver an extra car there. Although J.J. did not have a license and the three-hundred-mile trip passed through racially ambivalent territory, Jake trusted his son to drive there on his own. He needed the car; or, more important, he needed to give the boy a chance to demonstrate his ability to act responsibly.[5]

J.J.'s interest in cars soon graduated to a fascination with airplanes. He assembled intricate models and talked about becoming an aeronautical engineer. His father approved of the idea but, as always, suggested he set his sights higher. "If you prepare yourself," Simmons advised, "you can build and own your own aircraft plant."[6]

"We were raised with the idea that we were going to be bound for leadership," reflects Donald Simmons, now in his fifties and the youngest of the sons. "We were expected to do exceptional things. And in many cases, we were told there were things we couldn't do because we were his sons."

Donald remembers going with his high school football team to a "bootleg joint" (alcohol was technically illegal in Oklahoma until recently) after a game. "Hey, Simmons," a bartender yelled, pointing to the door. "You can't be here because your old man don't want you here."

Jake Simmons's scrutiny extended to all those living under his roof, including Blanche (Simmons's daughter by his first marriage) and several nieces who occasionally lived with the family.[7] John Simmons's daughter Johnnie Mae, a contemporary of Donald and Kenneth (the middle son), spent many summers with her uncle Jake, as well as most of her high school years, between 1946 and 1949. She says his verbal whippings were "like capital punishment—he would take you through your whole life." Then there was the vigilant inspection of her homework. "He was in constant contact with the teachers," she recalls. "They seemed to have a great togetherness."

For an attractive teenage girl like Johnnie Mae, Jake's watchful eyes extended well beyond the classroom. She enjoyed going to a popular burger joint on South Second Street called the Bright Spot, but when an "undesirable" element started going there he banned her from it. "You never know who you are looking at and you never know who is looking at you." he warned. "Just because you are away from home you might be tempted to do something and get away with it, but it doesn't always end up like that." The community, Johnnie Mae felt, "must have had a heartline in with him because if you were someplace [where] someone thought you shouldn't be he would hear of it before you got home."[8]

Although Simmons would not let his children become shoeshine boys, he made sure they understood the value of hard work. He paid them small sums to handle all the family's household chores and insisted that his sons also earn money on their own. During the

Depression, when J.J. was eleven, he asked his father to buy him a bicycle. Simmons told him he had to earn it himself.

"I was really a bit peeved," says J.J. "I couldn't understand, because I felt he had the money to buy it, and a few other kids' parents bought them bikes. But he wanted to encourage me to be an entrepreneur."

Curtis Publishing was offering a free bicycle to boys who could sell nine five-dollar subscriptions to the *Saturday Evening Post*. J.J. wrote away for the order forms. His father and a family friend bought the first two subscriptions, but it took six months to sell the rest. (Five dollars was a lot of money during the Depression.) When the bike finally came it was a big occasion: Simmons took his son to the Muskogee freight station to help him pick it up.[9]

Jake Simmons liked to see his boys selling magazines. When Donald was twelve, he and Ken, then fourteen, were taught sales techniques by their father to help them sell *Color* magazine door-to-door. (*Color,* one of the first national black magazines in America, was published in Charleston, West Virginia, and was an early competitor of *Ebony*.) Simmons liked the magazine, and believed that selling it would be a good way to teach his sons to deal with adults. "It was part of his plan to get us in a position where we would be comfortable with people, to not be afraid of looking them in the eye and telling them what you had to say," explains Donald.

The boys practiced their sales rap at home while their father listened and gave pointers. Donald, now a veteran international businessman and president of Simmons Royalty Company, responded eagerly to his father's advice and sold as many as two hundred magazines a month. Kenneth quit the subscription business early on, faced his father, and told him he was not going to sell anymore. (He never acquired a taste for business and is now a professor of urban planning and architecture at the University of California at Berkeley.)[10]

Donald enjoyed dabbling in the world of business, but he did not like cutting the lawn. Jake Simmons was a "stickler" (a popular word in the Simmons home) for having his lawn cut correctly. He expected his boys to edge the cut perfectly, to excel at lawn-cutting, as at everything else. By the time Donald was twelve, he came up with a better idea. "I figured it's a lot simpler getting other people to work for you than having to work for yourself," he remembers. Donald started a lawn service and got schoolmates to work for him. Charging homeowners double what he paid workers, he used the profits to pay his helpers to cut his own lawn. His father would not have it. He broke up his son's racket and forced Donald to cut the grass himself. The incident is the only confrontation Donald will speak of where time has not made him feel that the "old man" was right. He still defends his position as though the disagreement took place yesterday. "He had strong feelings about the dignity of work," says Donald. "I agreed with that. But I didn't agree that it meant you had to cut your own lawn."

Lawns were not the only things that had to be just so. Jake Simmons was a stickler for having everything done properly. That was, he believed, the only way a person with dignity should do anything. "You were always racially conscious in my home," says Donald Simmons. "He believed that the way for black people to compete in the world was that we could never be 'as good as'—we had to be better. Booker T. Washington made an indelible imprint on my father about that. He let us know that he expected anything we did to be better than anyone else. I resented it sometimes. If you cut the lawn, you had to do it better. If you cleaned the house, you had to do it better than anyone else. If you made an A, you had to make an A+, or ask if they gave A+'s. You didn't get any compliments from my father for being ordinary, or as good as."

Life with Father wasn't always easy. Soon after Johnnie Mae moved in with her uncle's family, she entered the segregated Manual Training High School and joined the school's marching band. The band was reputed to be the best in the state. Its well-drilled students played lively jazz tunes at football games, band competitions, and local parades. Johnnie Mae (she prefers to be called Johnnie) was part of the colorful baton-twirling group of majorettes who led the band. It was quite the thing to do, and Johnnie was honored to be one of the "high-steppers" who stood at the end of the line and kept the others on a straight path when they turned corners.

Johnnie will never forget Muskogee's 1947 Christmas parade. Before it began she had heard her uncle Jake complain to other parents that every year Manual Training's popular marching band was relegated to the very back. Soon after the majorettes started strutting their stuff, they passed Simmons's office on Second Street. The moment he spotted her he jumped from the sidelines and charged through the crowd, his face contorted in anger. "Come on out," he ordered, grabbing her arm. "You will not march behind all the white people." To this day, Johnnie does not know whether any of the other parents pulled their children out of the parade. "I was too embarrassed to ask because I didn't want them to know if they didn't already know I'd been pulled out," she says. "I wanted to just go someplace and hide. I did—I got lost in the crowd. It was one of the most embarrassing moments of my life."

Simmons had a similar response to Kenneth's high school choir. When he heard that a hotel where the choir was performing insisted that his son take the freight elevator, he forced him to quit.

Football did not pose the same problem. All three boys played during high school in a segregated statewide league. Muskogee loved football. High school games were major social events, especially in the black community. Jake and Eva were avid fans; they rarely missed one of their sons' games. J.J. III, six foot two and 230 pounds, played center for Manual Training High School's varsity team for three years (two of them championship seasons) and made all-state center twice in a row. Donald also played varsity for Manual Training for three years (including another championship season in 1950), and was all-state guard. Ken went on to play college ball at Harvard.[11]

Although Jake Simmons was repressive and strict at times, says Donald, often "he was a real softy. If we really wanted something we generally got it—if he was able to buy it. If he wasn't able to buy it you'd certainly never know he wasn't. He'd probably give you a reason why you shouldn't have it." As with any parent, persuasion depended on the right approach. With Simmons, this meant carrying off a good rap.

"To get the car on Sunday," remembers J.J. III, "I would have to sit through a lecture about how much responsibility there was to have a $2,500 automobile out there to carry girls around in. He really just wanted to hear me talk, to have me justify the need for it. He felt that the key to a man's success was the ability to talk and express yourself, which he did so well. I knew he was already going to let me use the car, but I would give him all kinds of rational arguments about how necessary it was for my development to do these kinds of things."[12]

The hardest thing about having Jake Simmons as a father, Blanche

Jamierson recalls, was that "you never got out of anything." As with her cousin Johnnie, Simmons was also responsible for Blanche's most awkward moments. Early one morning she tried to sleep late while her choral group had an engagement. When her music teacher appeared outside the house and honked the horn she pretended not to hear anything, figuring the group would leave her behind when they realized she was still in bed. The teacher entered the house and asked Blanche's father why she was not ready. Simmons rushed up the stairs and, in full view of the other choral members, "pushed me out the door, nightgown, dress and all, and I had to dress in the car."

Simmons was also strict when it came to Blanche's music lessons. He wanted her to be a concert pianist and expected her to practice regularly. (She had tried to take up the drums instead, but he thought it too unladylike.) Simmons's boys were also expected to play musical instruments. J.J. played the violin. Blanche remembers that they used to schedule their practices together, in nearby rooms, in the hope that the horrible noise would drown out their sluggish effort and inspire their father to cut short their rehearsal. Jake got wise to the conspiracy after a few ensemble sessions and forced them to practice separately.

By the time Kenneth and Donald (born in 1933 and 1935, respectively) were teenagers, both Jake and Blanche were out of the house and had started their own families. Kenneth and Donald were very different in character. They both got excellent grades, although Ken continued with his education longer than his more business-minded brother; Ken graduated from Harvard, then received an additional degree from Berkeley. Don excelled in public speaking. He toured Oklahoma and Texas as part of the school's expert debating team, and was so successful that white and black students across the state elected him as their first black "Youth and Government" governor in 1952. He spent a day handling the governor of Oklahoma's job and became the pride of the state's black lecture circuit.[13]

Donald was the more congenial of the two brothers. While Ken was quiet and reserved, Don was boisterous and fun—the kind of person who, as Johnnie Mae describes it, "could just sway in the wind." Blanche describes Ken as "kind of a loner." She says that he takes after J.J. in that both men are very serious and somewhat guarded. "When you talk to Don you kind of know what he's going to do and how he feels," Blanche explains. "I can never tell exactly how the other two feel." J.J. agrees that he is more solemn than Donald and less likely to express his emotions by doing something like crying during a tragic event.

Both boys were popular in different ways: Don was the "hell-raiser " and Ken was the cool ladies' man. They were among the only kids in the neighborhood with a "rumpus room" in the house. During Jake and Eva's frequent evenings away from home for business or civil rights dinners, the space became a party room. Friends would bring the latest jitterbug records for dancing and enter through the garage to avoid detection by neighbors (who reported everything they saw to their father). The boys' only obstacle was Eva's elderly grandmother, Annie Escoe. During the thirties and forties Escoe lived with the family. Although she often acted as babysitter Donald, ever the deal-maker, found ways to throw the parties anyway. She always went to bed at dusk and her eyesight was very poor, so he would quietly close all the shades in the house, then tell her it was nighttime. On those occasions when Escoe did manage to catch the boys committing punishable offenses Don used a different approach. "She'd like to chew tobacco and Uncle Jake didn't want her to," recalls Johnnie Mae. "Don would to the store and get her some tobacco while they [Jake and Eva] were gone, and when she'd start naming off the things she was going to tell them that we had been doing he would say that he was going to tell that she had been chewing tobacco. That would put a lid on all the things she was going to tell."

The family enjoyed Donald's humor. He liked to dress up in his father's most expensive clothes, then walk down the church aisle while the elder Simmons addressed the congregation. Kenneth and Johnnie Mae would watch gleefully from the audience to catch Jake's expression when he caught sight of his elegantly attired son. "He would always reprimand him," remembers Johnnie Mae, "but Don would take it with a grain of salt because Uncle Jake would just be there smiling down at everything."

Jake Simmons was acutely aware that his children looked upon him as a role model. He was careful to set an example of thrift and material conservatism. His interest in cars like the snazzy Pierce-Arrow faded with the lessons of the Depression. During the late thirties and forties, he drove Pontiacs and Plymouths, then switched over to Hudson Hornets—to him the ideal automobile. He wanted pragmatic medium-priced cars which held out after many long drives between Texas and Muskogee. Although he always lived comfortably (trading for a new car every few years), he believed that money, like people, needed to be properly employed to achieve its greatest potential.

Eva was the same way. Her husband tried to buy her expensive gifts but she shied away from most jewelry, limiting costly purchases to fine clothes, like the $400 Eisenberg suit he bought her during the fifties. "She was the soul of dignity," recalls Doris Montgomery, a family friend. "Completely without ostentation and never a braggart. She didn't have any need to flaunt her wealth—she was a self-sufficient person in that regard. I remember during the sixties, *Ebony* wanted to put him on a list of the wealthiest black men in the United States. Eva wouldn't have it. She said, 'That's too showy and cocky, people would be thinking we have more than we have.' "[14]

Simmons's behavior was anything but frivolous. His small talk, even with a stranger, was direct and to the point, and his deeds were done with a defined purpose. Money, too, had its uses, the highest of which was realized by giving it away. He approached giving—giving money, giving advice, giving time—with a fervor that can only be described as religious. "He believed that giving was a way of life," recalls the Reverend E. W. Dawkins, a squat, warm-natured man who was Simmons's minister for many years and now runs a church in Tulsa. Dawkins is a true Oklahoma minister. Quick to smile and an expert storyteller, he keeps a yard filled with tall mustard greens for members of the community to pick whenever they want. Recalls Dawkins, "I heard him say many times, 'You can't beat God's giving: the more you give, the more he gives to you.' That was his philosophy and he never got tired of it."

Faith in God meant a lot to Jake Simmons. He believed such faith was responsible for worldly advancement as well as spiritual peace. "If you don't have God," he told his family, "you don't have anything. You need God in your life to get anywhere."[15]

Simmons's faith was channeled through the African Methodist Episcopal Church, a Christian denomination that allowed the average practitioner an opportunity to give more, in the way of influencing and leading the congregation, than any other religion in the area.

The African Methodist Episcopal, or A.M.E., Church was tailor-made for someone like Simmons. It was founded in 1787 by Richard

Allen, a black freedman in Philadelphia. Allen was worshipping with a Methodist congregation which made him sit in the back of the church. During a service one day he attempted to walk up to the altar to pray and was turned away. Disgusted, he and other Methodist freedmen formed the A.M.E. Church, the first Christian denomination created by blacks.

Today there are about one and a half million A.M.E. Church members in the continental United States and another two million abroad. Laymen have a tremendous amount of power. Every four years each congregation elects delegates to attend a national convention where A.M.E. policy is shaped. Lay committees select the bishops who form the church's governing board. Members are not accountable to any pope, evangelist, dogma, or white leader. The A.M.E. Church was formed by blacks for blacks. Within it Jake Simmons found his spiritual home.[16]

Simmons's father had been a committed Baptist. Jake Jr. switched from the Baptist Church to the A.M.E. Church soon after he married Eva, who was a member. Their family joined the Ward Chapel A.M.E. Church, on Denison Street in downtown Muskogee. It became the center of their social universe. The Simmonses always sat in the last pew of the church so they could keep their children quiet. When Jake had something to tell the congregation—which was quite regularly—he would leave Eva in the back row and move up front. Near the end of each A.M.E. service, after the reading of Scripture and Psalms, the consecration hymn, and the preacher's sermon, the minister extends an invitation for people to join, or address, the congregation. "When you looked and saw him coming up and sitting on the front seat," remembers Dawkins, "you knew that Jake's going to talk this morning. That Jake—he could talk. And he'd do it often, when he had something on his mind that he wanted to get over. Anytime you saw him on the front seat, you knew."[17]

The subjects Simmons brought before his fellow churchmen varied. Often they were political. Sometimes he lectured on social or civil rights issues. "People would do what he said," says Dawkins. "They just had that kind of faith in Jake."

So did Dawkins. When the minister did not feel up to putting together a Sunday sermon, he would call Simmons the day before and ask him to write a speech. Simmons would knock one off in a half hour. "He was very enlightened on the Bible," remembers Dawkins. "He quoted Scripture right often."[18]

Church life was not a once-a-week affair for the family. Wednesday nights were prayer meetings, where members would sing together and discuss the Bible. Friday evenings were "class meetings." The church divided its members into groups of ten people. Simmons was a class leader. Each Friday he led his group in discussing what God did for them during the past week. After Sunday evening service Jake often invited his sons' friends home, where he would discuss world affairs, and have one recite, out loud, excerpts of a book he had read that week.

On a national level, Simmons became one of the A.M.E.'s most active laymen. From the late 1930s until his death he was the president of the trustee board of Ward Chapel Church, for which he served as a delegate during national meetings every four years. He was a major contributor and trustee of the church's Arkansas-based seminary, Shorter College, for forty years. He was also on the lay committee which selected the church's full-time ruling body of bishops.[19]

International politics was an important subject at the Simmons dinner table. Jake expected his children to keep up to date with news from the radio and the many magazines he subscribed to. He discussed most issues with them in the context of what they meant to the black race. He and Eva banned the popular radio show "Amos 'n' Andy" because it reinforced racist stereotypes. During the thirties he assigned J.J. to scan the newspaper each day and report to him after dinner on Japan's latest activities in its Manchurian war. "Dad was an early admirer of the Japanese," recalls J.J., a stalwart "Made in the U.S.A." booster who grimaces when he sees government officials driving foreign cars. "He predicted they were going to rule the world and by God, it's come to pass."

Jake Jr. identified with the Japanese people as a kindred dark-skinned people—according to J.J., "much to the dismay of the Japanese" he met. Simmons admired their traditional marital values, work ethic, and military acumen. Late in his life, when Japan had emerged as a dominant economic force, he would lecture young people to follow their example. "These Japanese are dedicated," he'd say. "And so thrifty!"

Simmons even found something to admire about Japanese imperialism during World War II. In 1942, when the Japanese drove the British from Singapore, he held a small celebration at home, telling his children, "The world's never going to be the same for white folks." Although he felt the Japanese would lose the war—and was proud, in 1944, when his son J.J. fought in the second invasion of the Philippines—Simmons regularly lectured his boys that the "war for democracy" would inspire independence movements, and lead to the emancipation of dark-skinned people worldwide. He despised fascism, and although he never flirted with the leftist ideology of men like Paul Robeson, he applied a racial analysis to his view of world politics. To Simmons it was simple logic that the ideals esposed by colonial powers in attacking fascist domination would be used to justify national liberation movements everywhere. "They'll be put out of India," he would say. "They'll be put out of Indonesia. Lastly, the winds of change will reach Africa, and our people will get their freedom."[20]

Although Eva rarely joined in her husband's political discussions she always accompanied him to social and political functions. Simmons family friend Doris Montgomery believes that Eva had "implicit confidence" and trust in her husband's decisions, yet was chiefly responsible for running the home. "She was his support," Montgomery says. "But she was not like him, Two people like Jake Simmons couldn't stay in the same house."

Granddaughter Donna recalls that Eva "spoke all her opinions straight out—she was never one of those wives that was scared of her husband. He didn't treat her like a child, like a lot of men in that generation treated their wives. She did the cooking and cleaning, but it was not as though it was subservient—it was just what she did, and he treated her as an equal."

The Simmonses, of course, had their share of arguments. The most divisive issue was Jake's persistent habit of offering dinner to anyone who set foot in the house. J.J. says that this bothered his mother because she did not always think the food she had prepared was suitable to serve to guests. "But the old man thought that if it was good enough for him it was good enough for anybody," J.J. recalls. "Dad was forever bringing people in unannounced and boy, she didn't like that. But you'd never know it. They didn't express those kind of things outwardly among us. They'd make sure it'd be behind closed doors. She didn't want his authority questioned."

Jake and Eva's arguments rarely got out of hand; relatives do not recall them ever shouting at each other. Almetta Carter, Simmons's secretary for two decades, recalls that he often counseled her on marital relations. "Whenever you leave home," he'd advise, "make sure you make up with your spouse. Because you never know whether you'll make it back or not. You never know—you could get struck by lightning—then he'd feel guilty forever. So I always kiss and make up before I leave home."[21]

Politics was one of Simmons's great passions. Because most African-American leaders in eastern Oklahoma were perceived to be pawns of white benefactors, many blacks by the mid-1940s began to look to Simmons for independent leadership. Simmons was so self-sufficient that he owned the two-story office building he worked in, on North Second Street. He rented space to some of the most important black professionals in the city, including a doctor, dentist, and real estate broker. Between the people who walked in and out of these offices and those who patronized the pharmacy on the ground floor, few blacks in Muskogee had any trouble knowing where they could find the outspoken oilman. In 1946, at the age of forty-five, he made his first bid for public office.[22]

The 1946 citywide election in Muskogee presented an important opportunity for Simmons because ten City Council seats were up for grabs. Members of the sixteen-person council worked with the mayor to run Muskogee, and candidates usually ran on the same ticket as one of the leading mayoral candidates.

Two major Democratic tickets were contending for power in the 1946 election; the "Greater Muskogee" slate and the "Good Government" slate. Fifteen of the nineteen council candidates aligned themselves with one of the two tickets. At a time when a black candidacy was still unheard of in Oklahoma, both local parties refused to allow Simmons to join their ticket, so he became one of four "independent" candidates.

Simmons knew that running as an independent in Muskogee was a shot in the dark: the last time a black man in the state had been elected for any major office outside an all-black community had been in 1908. But he believed that blacks needed to run for office—if not to win, then to serve as an example to others that the electoral process could include them.

Local elections in Oklahoma were important community affairs. The ticket candidates did little individual campaigning, relying on the success of their heavily promoted slate to sweep them into office. Independents like Simmons had to maintain high profiles in the community, speaking at churches, street-corner "stump" rallies, and outdoor barbecues—his favorite speechmaking event. Before the days of television, political gatherings were the community's greatest source of entertainment. The soft drinks, beer, and barbecued meats were free, and the speakers knew how to stir the crowd. Everyone turned out. "If you throw a barbecue," Simmons used to advise his sons, "you can get people together to listen to anything."

Media advertising was rare for local elections during the forties. The most common form of publicity was (and still is in rural Oklahoma) colorful signs, posted on supporters' fences, stuck into the ground on front lawns, and set up along roadsides. The greatest expense of any election was the money which had to be paid to the city's precinct bosses to "get out the vote" on election day—hundreds of dollars which went ostensibly for transportation.[23]

In announcing the names of all nineteen council candidates the day of the election, the *Muskogee Phoenix* described the occupations of all the contenders. One was a used-furniture salesman, another was a plumber, another an owner of an oil company. Although he, too, owned an oil company, Simmons was simply listed as a "Negro leader."[24]

The electoral process in Muskogee allowed all citizens to vote for candidates from every district, making it impossible for blacks to elect someone the majority of white voters did not approve of. On April 2, 1946, Simmons received 1,059 votes, considerably more than any of the other independent candidates. Although he fell far short of the "ticket" candidates he trounced his opponents in the black districts, where voter turnout was unusually high. Within the city's five mostly black precincts he took virtually every African-American ballot, winning nearly three times as many votes as his nearest competitor.

Although he didn't win a seat on the city council he demonstrated, for the first time, the level of his political support.[25]

For every black vote Simmons won in Muskogee there were more than a dozen more African-American voters in other parts of eastern Oklahoma looking for a leader. Although state politics interested Simmons he was too realistic to run for statewide office. If he couldn't win a race in Muskogee, where one-quarter of the voters were black and he was well known, what chance did he have statewide, where less than one-tenth the voters were black?

Instead, Simmons took his impressive public-speaking ability and his newly demonstrated vote-winning power and went to work influencing the 1946 statewide elections. He became a manager of white oilman Fred McDuff's gubernatorial campaign, traveling around the state and setting up a network of support centers on his behalf. In the wake of the Supreme Court decision outlawing the state's grandfather clause, Oklahoma's black voters had become, to some politicians, a constituency worthy of attention. It was necessary, as one civil rights leader of the period termed it, for candidates to "latch on to some prominent black person to introduce him in the black community." Simmons believed McDuff was the most "liberal" candidate in a field of eight. McDuff's stand on civil rights was relatively enlightened, he was running against the "good ole boy," politics-as-usual gang, and he favored repeal of the state prohibition amendment.

During the campaign, an incident occurred in Oklahoma City which, more than any other, became publicly identified with the name of Jake Simmons. On June 6 Simmons set up a campaign stop in the center of Oklahoma City's black business section. The appearance, at the corner of Stiles and Second streets, was well publicized, and some 250 African-American voters gathered to hear Simmons and McDuff.

Simmons gave a welcoming speech, introducing his candidate. McDuff addressed the crowd. To warm up for a night on the campaign trail he had earlier exercised a personal protest against prohibition. Simmons later recalled that McDuff was very drunk when he climbed onto the podium. His words flowed loosely as he outlined the issues, speaking in the familiar and easy style of an Oklahoma politician. In the midst of his speech, perhaps warmed by their response,

McDuff leaned over to his audience and declared, "I have always loved good old southern darkies."

A murmur ran through the crowd. Simmons, who stood in front of the platform, leapt onto it, interrupted the speech, and whispered heatedly in McDuff's ear.

"I am not used to making apologies to anyone," a startled McDuff admitted to the audience. "But I'm told that Negroes do not like to be referred to as darkies. I meant no reflection, but only used 'darky' as an endearing term."

Most of the audience did not stick around to hear McDuff's concluding remarks. Another meeting was scheduled later that evening, but Simmons refused to go. He quit on the spot.

The next morning he walked brusquely into the office of Oklahoma City's *Black Dispatch* and asked editor Roscoe Dunjee to print his letter of resignation, which appeared in the next issue of both the *Dispatch* and the *Oklahoma Eagle* (Tulsa's black paper). The story was the lead article in both papers, headlined, JAKE SIMMONS QUITS FRED MCDUFF in the *Dispatch*, and MCDUFF CALLS HIM DARKEY; CAMPAIGN MANAGER QUITS in the *Eagle*.

Simmons's letter described the previous evening's incident, then continued:

> I am writing this letter for the purpose of informing Fred McDuff, H. C. Jones, Dixie Gilmer, Johnson Hill, Bill Coe [the other gubernatorial candidates], all white people, and all Uncle Tom Negroes who have not got the intestinal fortitude to resent this insult that the word Southern Darkey is no "Endearing Term," but is an Antebellum expression imputing inferiority to Negroes. I am calling upon all white citizens and many Negroes of this state to delete such words from their vocabularies. We are all Americans, and to be real Americans, we must have fortitude, and courage to not compromise such insults, but demand justice and equality for the minority group.[26]

Simmons threw his support behind a slightly less liberal candidate, H. C. Jones, who was popular with other black leaders. McDuff, once considered a contender, fared miserably in the next month's election. He came in fifth place, winning only 7,854 votes, compared to 138,000 for the top candidate. (Jones captured 79,273 votes.)[27]

In 1948 Simmons again became involved in the state's democratic primary election. This time it was a local man, county attorney candidate Ed Edmondson, who aroused the activist's support. And unlike the McDuff fiasco, Simmons's endorsement was to prove well placed. His bond with the Edmondson family would be one of the most important political alliances he would ever make.

If any clan can be called the ruling family of Muskogee politics it is the Edmondsons. Ed Edmondson, Sr., was a county commissioner in Jake Simmons, Sr.'s, ranching days. His son Ed Jr. became county attorney at twenty-nine and was then elected to the first of ten consecutive terms as a U.S. congressman in 1953. His other son, Howard, followed in his older brother's footsteps, starting out as county attorney and quickly becoming the youngest governor in Oklahoma history. He later served as a U.S. senator as well. The most recent generation of Edmondsons are also politically active; one is currently Muskogee's district attorney and another is a district judge.[28]

Ed Edmondson, Jr.'s, father knew Jake Simmons, Jr.'s, father well. As Haskell's most prominent black rancher, Jake Sr. was, as Ed Jr. recalls, "somebody that needed to be talked to." The senior Edmondson exposed his children to politics by bringing them along when he visited places like the Simmons ranch. Ed Jr. remembers Jake Sr. discussing politics on the front porch with his father during the thirties. As they drove off, Ed Sr. told his young son, "That's the kind of man who, if he tells you he's for you, is for you."

Jake Simmons, Jr., was also a supporter of the senior Edmondson, but it was Ed Jr.'s race for county attorney which brought him to the speaking platform. Young Edmondson had impressed him early in 1948, through his instrumental role in organizing a civic assembly to honor the Freedom Train (which traveled across the country bringing important historical documents like the Bill of Rights to middle America). The gathering attracted many of Muskogee's most important citizens to city hall. Simmons was pleased to find the seating integrated, and both black and white choral groups appearing on stage. There was even a moment when the entire audience joined hands—an unusual event in a segregated society.

Jake Simmons, Jr., also disliked the way Edmondson's opponent in the race, incumbent county attorney Chester "Cornbread" Norman, meted out justice. When a black criminal's victim was another black, he was let off easily. When the victim was white life and near-life sentences were the norm. "He was incensed that penalties were not uniformly imposed," recalls Edmondson. "He felt there were two standards of punishment."

Simmons campaigned for Edmondson throughout the black community. The candidate needed all the help he could get. Even by Oklahoma standards, where political races can resemble hillbilly blood feuds, the 1948 Edmondson/Norman contest was unusually bitter. "Cornbread" was well entrenched, and commanded immense influence. Edmondson ran as the "independent candidate" against Norman's "political machine." Two days before the decisive runoff election, on July 25, 1948, Chester Norman's campaign bought a large ad in the *Muskogee Phoenix*. Entitled "Politics Make Strange Bedfellows," it was a satirical cartoon of Edmondson (identified as "Little Ed") pulling a lazy bed filled with prominent supporters toward a finish line. "Bah jove!" the Edmondson figure remarks, "I thought I was a cinch to win with all you boys behind me. Quick, promise some more assistants jobs!" Waiting near the "July 27 primary" finish line stands a racist caricature of Jake Simmons, Jr. The figure is dressed in a zoot suit and wide-brimmed hat. Unlike the sketches of the other Muskogeeans the caricature bears no resemblance to Simmons, but is instead a stereotypical black hustler with a stupid expression and fat white lips. His hands are outstretched and he says, "I'm glad you and Bates ain't gonna pass me by, too. I got influence!" The figure of Norman is rushing past Simmons and the others, about to pass the finish line, remarking, "And *they* talk about machines."[29]

The election was close. Edmondson won by 720 votes out of 13,600 cast. His greatest plurality was in the black precincts. His advantage in just six of these precincts (out of a county total of sixty-six) amounted to more than 750 votes. Without the black vote, Edmondson might never have gotten his start as a politician. "Jake made a critical difference in my 1948 election," he recalls. "And I think he contributed to every campaign I was ever in. He was one of the most able supporters and friends that I had. He was the single most influential black leader in eastern Oklahoma. I think his influence and importance probably grew as you moved out of Muskogee County into other counties."[30]

Simmons's association with the Edmondsons eventually allowed him to pressure them to vote for civil rights legislation when they got to Washington. During the late forties, however, most advances in the struggle for racial equality came through federal court rulings. The NAACP, its treasury swelling with the contributions of World War II veterans, became increasingly determined to press for an end to segregation. In November 1945, Simmons and hundreds of delegates met at a conference of the state branches and voted to intensify their fight for educational equality. As head of Muskogee's local NAACP and president of the Negro Chamber of Commerce, Simmons took a lead role in raising funds for what became two of the most influential cases leading up to the Supreme Court's landmark *Brown* v. *Board of Education* desegregation ruling in 1954: *Sipuel* v. *University of Oklahoma* and *McLaurin* v. *University of Oklahoma*.[31]

Education was a critical issue to Simmons. He agitated loudly for it, especially in his home district. He applied pressure to the Muskogee school board to allocate funding for the area's black schools, and slowly won limited support. Through the 1940s a couple of small Negro elementary schools in rural Muskogee County were added, but it was not until early 1948 that the county commissioners agreed to address seriously the decrepit state of their black schools. They set aside $1.25 million in capital funds, part of it for a new black high school, and held a public referendum to allow a bond issue to finance it. In January 1948 (the first month of Ed Edmondson's term) the referendum passed. Black parents like Simmons, who had two sons and a niece in school, were delighted that $805,000 of the money had been earmarked for black schools. Donald was thirteen and Kenneth and Johnnie Mae were about fifteen. All were at an age where they could have taken advantage of a new high school—provided it was built promptly.

It was not. A year after authorizing funding the county had not spent a cent for the new school. The county commissioners and the Muskogee school board crossed swords over which group would have the right to let the contracts. The conflict pitted two regionally powerful groups against each other. The city's Board of Education insisted that an influential local architect named Frederick E. Zaroor be given the project. The county had different ideas, and withheld funding. City politicians tied the project up in court, lost, then got the Oklahoma legislature to pass a special bill transferring authority to build schools from the county to the local school districts.

The war over contracting powers continued, and it seemed that the new school—which the white community was not enthusiastic about building in the first place—might be mired down in litigation and political infighting indefinitely. Determined to get the ball rolling, Simmons took the case to county court and sued both the school board and the county commissioners. Purporting to represent himself and other similarly situated citizens, he fired off a double-barreled lawsuit, charging the county commissioners with being slow in building the school, and the city school board with illegally interfering with the commissioners' efforts. "By reason of such failure and delay in the commencement of said building for said Separate Schools," Simmons argued, "the Negro citizens, patrons, and children have suffered and will continue to suffer great inconvenience, injury and damages, resulting from insufficient educational facilities and qualifications, as provided by the laws of Oklahoma and will continue to suffer such injury and damages, unless relief is granted by this Court."[32]

Simmons also argued that the new state bill granting the school board contract-letting authority was unconstitutional. A county judge immediately granted him a temporary restraining order, stopping the school board from hindering the commissioners. The already complex battle became even more complex. On March 19, the school board responded that it had consistently tried to speed, not hinder, the construction of the school. The board also did its best to discredit its long-standing antagonist. Their response questioned Simmons's true motives and claimed, "The defendant denies that the plaintiff [Simmons] represents anyone but himself, and the plaintiff is not the real party in interest."[33]

Simmons's action did, in a sense, benefit the county commissioners, but he was certainly representing the black community at large. He was also the real "party in interest"—and not just because of his sons. Simmons had a score to settle with architect Fred Zaroor, and the lawsuit was probably a good way to do it.

Simmons's battle with Zaroor is a good example of the kind of racial combat he regularly engendered in Muskogee. Zaroor, now eighty-three and in semi-retirement, today says that blacks are "as nice a race as you'd ever want to meet—if they're educated," and that desegregation was "the greatest thing that ever happened to the country." Yet his hostility toward Simmons seems rooted in an unwillingness to tolerate an African-American conducting himself with defiant self-assurance. Zaroor remembers Simmons as "an egotistical cuss that nobody wanted to associate with . . . he didn't use any finesse to talk with people—he was more on the offensive side than the defensive side." What's more, continues Zaroor, "I don't think colored people ever felt that Jake was favoring them. He was favoring himself. As far as colored people were concerned, I don't think he had any idea of becoming associated with them. He wouldn't let a colored man walk into his house. I know. I've heard that time and time again."

Given the dozens of black Oklahomans I interviewed who had indeed visited the Simmons home the outrageousness of this comment colors Zaroor's recollection of how the two men came to be enemies. According to Zaroor, Simmons approached him and asked that the new school be named after him. The architect says he refused even to consider the idea, and told Simmons so one day when the two men met by chance in downtown Muskogee. Simmons, if one is to believe Zaroor, got heated up and "cussed me out." Then, recalls the architect, "I just plain told him: I said, 'I won't take that off a nigger.' That really got him; now you don't say those things. He said, 'You haven't heard the last of it,' and walked off."

Simmons's lawsuit was withdrawn after a few months, and the board of education ended up with the power to build the new Manual Training High School. It was, however, built quickly. Not surprisingly, Fred Zaroor was its architect.

When it came to politics Simmons was, above all, his own man. His independent livelihood allowed him to develop his own racial litmus test before deciding whom to support. When Johnston Murray approached him in 1950 to endorse his candidacy for governor, Simmons's litmus test was particularly tough.

Johnston Murray was the son of "Alfalfa Bill" Murray, a notorious racist who served as governor between 1931 and 1935. While in office Alfalfa Bill did everything he could to suppress blacks. He took the low road of folksy populism, berating college-educated city slickers and refusing to let technicalities like the law interfere with his religiously fanatical bigotry. In the early 1930s *Black Dispatch* editor Roscoe Dunjee challenged Governor Murray in the state supreme court and won a landmark case for residential desegregation. During the conflict Murray called a group of blacks to the capitol. When they arrived, he shouted, "You niggers come in here. Do you see this line I have drawn? I want you niggers to stay on the south side of that line and the white folks must stay on the north side. I do not have the law to enforce this edict BUT I HAVE THE POWER."[34]

Alfalfa Bill's views grew even more extreme as he got older. In 1948, he published a book—on his own printing press—called *The Negro's Place in Call of Race; The Last Word on Segregation of Races Considered in Every Capable Light as Disclosed by Experience*. The ninety-seven-page diatribe, more than half of which was devoted to anti-Semitic appendices, stressed the need for Aryan people to "teach their children to beware of crossing with any kinky or curly hair races." The main thrust of the tract was that racial mixing, promoted by an insidious Jewish Communist conspiracy, posed a grave danger to America. As evidence he presented a chart titled "Dr. Hunt's Brain Weights," purported to be the findings of a doctor who weighed the brains of dead soldiers during the Civil War. The brain weight of forty-seven white soldiers was said to average 1,424 grams. "Pure Negro" brains averaged 1,331 grams; half-white, 1,334; one-eighth white, 1,308; and one-sixteenth, 1,280 grams. How the mysterious Dr. Hunt came to assess so accurately the corpses' racial heritage is never mentioned. Murray was nonetheless impressed by the statistics. "These are startling but stubborn facts," he wrote, "and should alert every American, black and white, to stand foursquare against the Communists' and yellow Negroid-Mongrels' efforts to create more of the last named. They should ever remember the Mongrels—after half blood—lose brain power as the white blood increases, so at last [have] just enough brain to make them impudent, which they are, always."[35]

Johnston Murray was not nearly as objectionable as his father. In fact, most found him to be a congenial, good-hearted man. His campaign slogan, which was plastered on billboards across the state, was "Just Plain Folks." He appeared in ads without a tie or jacket, a rough-faced, hairy-armed man rolling up the sleeves of an open white shirt. The ad copy in a black newspaper neglected to mention a single campaign issue. It read, "OUR SLEEVES ARE STILL ROLLED UP . . . and they are going to stay rolled up. All of us plain folks are well aware of the big job that lies ahead . . . HELP ELECT A GOVERNOR FOR ALL THE PEOPLE."[36]

Murray needed black support for his Democratic primary runoff election against William O. Coe in July 1950. A third candidate more popular among blacks was knocked off the ballot, so Murray had little trouble lining up a few important black leaders, promising them patronage jobs and, in all probability, distributing the necessary cash.

Coe was also eager to gain the black vote. His appeal was far more sophisticated than Murray's. Photographed in a dark suit and tie, looking like a thoughtful young man who knew the score, Coe's black newspaper ads listed a series of campaign issues. He accused Murray of having sold out to "moneyed interests," in return for a promise not to institute social reforms. Coe promised these reforms, and specifically promised blacks he would be "a governor for the rank and file, without regard to race or creed."[37]

What set prominent blacks like Simmons against Murray was that during campaign speeches he refused to say that he would treat racial issues any differently than Alfalfa Bill had. Instead, he regularly told his audiences that if elected he would be guided by the philosophy of his father.[38]

Johnston Murray first visited Jake Simmons at his office in mid-1950 to win his support in the primary. Simmons said he would not hold the politician's ancestry against him. He offered to back Murray—on one condition: that Johnston Murray publicly disavow his father's comments about blacks.

Because Alfalfa Bill's notoriety elevated his son in many circles, Johnston steadfastly refused to disavow him. This, to Simmons, was intolerable.

In a Democratic-controlled state the party primary determined who would be the next governor. William Coe came in way ahead of Murray but did not win a clear majority. A runoff was called, and Murray pulled out the stops to catch up.

Johnston Murray swung around eastern Oklahoma to once again try to win the black vote. At the state's black institutions in Taft and Boley, patronage bosses "Major" H. C. McCormick and Eddie Warrior—eager to hold on to their positions—spoke out for Murray. But Murray still had the problem of Jake Simmons. The Muskogee leader had already started publicly criticizing him. If Simmons could be neutralized, thought Murray, the black leaders in his camp would be able to sway the region's African-American vote.

Murray and an aide paid Jake Simmons a visit. Their campaign was well financed. According to Donald Simmons, Murray told his father, "I respect the reasons you couldn't support me and I don't expect you to come out for me in the runoff. But the black vote could determine the difference in this race, so why not take an out-of-state vacation for three weeks until the runoff, and I'll give you $10,000."[39]

Murray's assistant had a briefcase filled with money. Simmons turned him down. Although Murray had reportedly bought up all available copies of Alfalfa Bill's embarrassing *Negro's Place* book, Simmons already had one. For the next three weeks he quoted from Alfalfa Bill's hard-to-find book as he campaigned for Coe in black communities across the state. "Daddy brought the book out on the platform and really worked him over, every chance he got," Donald recalls.

The election was closer than anyone had imagined. Simmons, true to form, delivered to Coe roughly two-thirds of the black votes in Muskogee. It was not enough. Murray won the gubernatorial race by a slim 886 votes. "They stole the election from me," Coe declared. but he never proved it.[40]

Simmons hoped that his effectiveness in the gubernatorial election would set an example which his oldest son, J.J. III, would follow. After being away for seven years J.J. returned to Muskogee in 1949 at the age of twenty-four. He had spent a few years at the University of Detroit, served in the war, and completed a bachelor of science degree in geological engineering at St. Louis University. America's first accredited black geological engineer, he went into business with his father as secretary-treasurer and vice president of the Simmons Royalty Company.[41]

J.J. wasted no time in becoming an important figure in local politics. He and his mother helped run the Muskogee chapter of the NAACP, allowing his father time to spend more time managing the organization on a statewide level. His first major achievement came in August 1954, when he successfully pressured the mayor and city attorney to desegregate all of Muskogee's public recreational facilities, including its swimming pools. J.J. used a similar argument on a statewide level: that there were no equal facilities for black citizens, and that the government was legally bound to create them or desegregate. He appealed to Governor Raymond Gary to open the state's five multimillion-dollar lakeside resorts to blacks. Gary found little fault in Simmons's argument. He quickly desegregated the state lodges by executive decree.

J.J.'s actions outraged the racist element in his community. Just as his father had been years earlier, he and his family were besieged by threatening phone calls and written death threats from the Ku Klux Klan. Instead of backing down he turned the letters over to the FBI and responded the way his father taught him to. "I went to Sears and bought my first automatic shotgun," he recalls, "which I still own today."

Attempts at economic pressure were as ineffective with J.J. as they had been on Jake Jr. In retrospect, the younger Simmons believes he persevered because he did not have to work directly for any white-owned business. "I enjoyed the economic independence of my association with my father," he says, "because they couldn't do anything with us. I didn't have any fear because I didn't hold any job."

J.J.'s struggle for racial dignity was fought with the same "bulldog tenacity" (one of his father's favorite phrases) as Jake Jr's. B. F. Rummerfield, founder and chief executive officer of Tulsa's Geodata Corporation (one of the nation's largest geological data companies), befriended J.J. early in his career. He says that in the mid-1950s the black geologist asked him to sponsor his membership in the important American Association of Petroleum Geologists (AAPG). Rummerfield, who was an active member, told him that the organization had an unwritten law to exclude blacks. "But how do you feel about it?" Simmons asked. "The color of your skin makes no damned difference to me," Rummerfield said, signing the sponsorship form. The application did not mention race. The association relied upon the discreet racism of its membership for compliance, and never knew what it was doing when it admitted Jake Simmons III as its first black member in 1958.

Several months later, at the association's annual convention in Dallas, Rummerfield's covert action was discovered. Texas law forbade a black person from staying in the same hotel as a white. J.J. checked into the convention at the luxurious Adolphus Hotel and insisted on registering for a room with his wife. AAPG officers argued and cited local law, but he would not budge. Finally a panicked Shell Oil executive telephoned Rummerfield, Simmons's sponsor.

"You know him!" the oil executive cried. "You talk to him. He's creating a big problem!"

Simmons came to the phone. "Why do you want to stay there?" Rummerfield asked.

"Because they won't let me," Simmons said. "I have to make a point."

After threatening to call the police, the hotel staff and association officers relented, and Simmons was given a particularly nice room. Rummerfield remembers that he was later contacted by an indignant AAPG member from the convention who had been shocked by the white man's sponsorship and complained, "But he's a nigger—he's black!" To which Rummerfield replied, "That has nothing to do with his mental capacity. His family history would put ours to shame."[42]

J.J.'s activism spilled over into his religious life. But his church was no longer the church of his parents. In 1945, while attending the University of Detroit, a Jesuit institution, he became so impressed by Catholicism that he converted.

Although Eva Simmons had difficulty accepting her son's religious conversion Jake Jr. welcomed it. "He encouraged me," recalls J.J. III. "He had a lot of vision and understanding, and he wanted to support us in whatever we did. Dad always thought it was necessary for you to be whatever you were in order to be sure that you worshiped God in some capacity. And he liked the regimentation and discipline that was taught in the Catholic Church." (Simmons's feelings about his son's faith would prove prophetic. J.J. was the only one of the three boys who converted from his father's denomination. Today he is the most religiously observant of the three, attending church every Sunday.)

In 1953 J.J. and his wife, Bernice, who was also Catholic, decided to send their children to a parochial school. When they tried to enroll them at Sacred Heart, the local Catholic school, they encountered the resistance of a priest who served as principal. Although it was a year before the Supreme Court school desegregation ruling (which did not apply to private schools), Simmons was morally outraged. He appealed to a higher authority—the regional bishop, in Oklahoma City—and convinced him to allow his oldest son, Jake IV, into Sacred Heart. The young boy, who is now a lieutenant colonel in the army,

:y east Texas oil field.
ıerican Petroleum
:tute)

skogee, "Queen City of the Southwest," at the turn of the century.
'sa University, Special Collections Library)

Black-Indian dance at Fort Gibson, Indian Territory, while blacks awaited enrollment for allotment of Creek tribal lands, about 1904. Every man, woman, and child received 160 acres; it was the largest land disbursement to former slaves in North American history. (*Archives and Manuscripts Division, Oklahoma Historical Society*)

(LEFT) Jake Jr.'s mother, Rose Simmons, with her youngest child, Arvada approximately 1910. Raised on the ranch of her grandfather, Cow Tom, the Creek tribe's only black chief, Rose was a tough, sharp-eyed frontier woman. "Training" children for thrift and discipline was her greatest joy. (*Simmons Family Collection*)
(RIGHT) Jake Sr. and Rose Simmons with granddaughter Ophelia, whom they raised after her mother was sent to prison for killing her husband's mistress, photo taken approximately 1920. (*Simmons Family Collection*)

Jake Jr. with his first son, J. J. III, about 1926. (*Simmons Family Collection*)

The Simmons family, about 1940: (left to right) J. J. III, Donald, Eva, Kenneth, and Jake Jr. The east Texas oil boom provided the family with a comfortable living, but taking the school board to the Supreme Court in an equal education lawsuit made Simmons the most controversial African-American in the state. (*Simmons Family Collection*)

Tuskegee Institute's 1918 championship basketball team, with Jake Jr. in the center, holding the ball. Even as a teenager he knew how to stand out from the crowd. (*Collection of Blanche Jamierson*)

J. J. Simmons III, about 1930. (*Collection of J. J. Simmons III*)

As a high school student in 1952, Donald Simmons was the first black elected "youth governor" of the Oklahoma YMCA's Youth in Government program. (*Simmons Family Collection*)

ake Jr. in Lagos, Nigeria, holding granddaughter Annamarie, mid-1960s. (*Simmons Family Collection*)

Jake Jr. and J. J. III out in the oilfield, about 1955. (*Simmons Family Collection*)

ake Jr. in Accra, Ghana, with associate Jack Zarrow and Ghanaian fficials, approximately 1976. (*Simmons Family Collection*)

Donald Simmons, executive director of the Harlem Commonwealth Council, 1971. (*Simmons Family Collection*)

Current Interstate Commerce Commissioner J. J. Simmons III (right) and Donald Simmons with then Vice-President George Bush, 1986. Although they are Democrats, members of the current generation, like their father, straddle the political fence. (*David Valdez, The White House*)

Almetta Carter, secretary of the Simmons Royalty Company for twenty-five years. During the mid-sixties, the highly qualified and articulate Carter was Jake Simmons's secret weapon to integrate segregated Muskogee businesses: he would send her out to apply for jobs, then raise hell if she was discriminated against. (*Jonathan Greenberg*)

Sylvan Amegashie, former Ghanaian oil minister, in his London office. For his role as "mentor" to Amegashie and subsequent Ghanaian officials, the nation awarded Jake Simmons the Grand Medal—its highest honor—in 1978. (*Jonathan Greenberg*)

Edwin Van den Bark in 1986. He first commissioned Jake Simmons to work for Phillips Petroleum in Africa in 1964. Now retired, Van den Bark ran Phillips's worldwide exploration program for more than two decades and established a reputation as one of America's greatest international oil finders. (*Jonathan Greenberg*)

Jake Jr.'s only surviving sibling, John Simmons, in 1986 with his wife, Daisy, on the Oklahoma city street that bears his name. (*Jonathan Greenberg*)

Jake Simmons's old office building, at 282 North 2nd Street in downtown Muskogee, was the last major black-owned establishment on a street that once boasted one of the largest African-American business districts in the Southwest. The building was demolished by the state in 1986 to make way for a parking lot. (*Jonathan Greenberg*)

withstood considerable abuse from the school administration and his classmates. Eventually the resistance subsided, and all four of J.J.'s children attended the school. Simmons also single-handedly desegregated the Muskogee branch of the Knights of Columbus, the Catholic fraternal organization, in 1955.[43]

Integrating himself into his father's business was almost as difficult. Jake Jr., like most independent brokers, was something of a loner. He did not like having to answer to anyone—boss or partner. A son, of course, was different. J.J.'s geological education provided the expertise necessary to expand the Simmons Royalty Company beyond his father's royalty business. Jake Jr. had brokered and invested in scores of wells through the years (many of them dry holes), but it was not until his son joined the company that he initiated his own drilling operation.

The deal which inspired Jake Simmons to start drilling on his own came about through quick action and a bit of luck. In 1949, near the tiny town of Boynton, Oklahoma, a pumper was observing the drilling of a well for Midcontinent Oil & Gas, one of Tulsa's largest companies, when he struck oil. Although it was 10:00 P.M., he had orders to inform one of the company's landmen of a commercial strike whenever it occurred. (This was so the agents at headquarters would be able to snap up drilling leases on surrounding land before any of their competitors found out they were valuable.) The pumper rushed to a local phone and called Tulsa, never noticing that he was on a party line. Willie Harrison, a childhood friend of Jake Simmons's who shared the party line, happened to pick up the phone during the conversation. Instead of interrupting he listened in. When the men got off the line Harrison called his old buddy the oilman.

Although he was already in bed Simmons threw on his clothes and rushed out to Harrison's place. As the sun rose at 5:00 A.M. the next morning the property owners whose farms adjoined Midcontinent's wells had an unexpected visitor. By the time Midcontinent's landmen got to the area Simmons was at the county clerk's office, registering his claim.

Simmons financed a promising first well on the Boynton land himself, but further exploration promised to be very expensive. Simmons wanted to drill eleven holes. Always eager to spread financial risk, he called his friend B. F. Phillips to ask if the Texas oilman wanted to invest in a working interest in the wells. They struck a deal over the telephone.

"When do you need the money?" Phillips asked.

"In the morning," said Simmons.

"I'll wire it to you," Phillips said. "You send the papers down here. I'll sign them later." The next day, with nothing but Simmons's word, Phillips reportedly sent more than $100,000 for his stake in the deal. The wells brought in tens of thousands of dollars' worth of oil a month. It was Simmons's first successful production operation.[44]

On that project and subsequent exploration deals the Simmons Royalty Company generally farmed out their drilling operations to independent contractors. Jake Jr. consistently minimized financial risk by finding investment partners. Usually he would have to pay for the lease and first well himself. With luck and initial production figures he would convince investors to buy "working interests" on a property. His optimum goal, though seldom realized, was to retain a 50 percent interest (after the landowner's standard one-eighth royalty) while raising enough cash from selling a few 10 percent or twenty-five percent stakes to reimburse his original expenditure and finance further drilling.

Although oil exploration is a high-risk enterprise, Simmons did his best to make investors feel otherwise. Those who did not turn a profit sometimes resented him for this. It was not mere hype. Simmons was what his son J.J. calls a "complete optimist." He truly believed in the high expectations he used to promote his prospects. J.J. countered this with the pessimism of a professional geologist. He recalls that only about half his former company's investors ever made their money back (a figure, he hastens to add, which exceeds the national average).

Among Simmons's early investors was the young Howard Edmondson, county attorney Ed Edmondson's brother and a future governor. "He talked my brother into pitching about $1,500 into the only wildcat he ever went in on" Ed Edmondson recalls with a laugh. "Jake said it was bound to be a gusher, but he was wrong on that one. It took Howard two years to pay off the note."

Simmons also worked out deals with some of his drilling contractors, offering percentages and even loans in return for services. In

1951 when a contractor which owed him about $10,000 was unable to make good, Simmons was given ownership of its old drilling rig. The equipment gave his company an opportunity to do its own drilling, but not until they had invested roughly $20,000 to have it rebuilt at the Muskogee Ironworks.

With the addition of the rig Simmons Royalty became a better-rounded operation. While his father ran the office and brokered deals, J.J. ran field operations, arriving at the cable tool rig at 6:00 A.M. and returning home at dark. While drilling a well, the Simmonses would have three shifts of two men—the driller and the tool pusher—working around the clock. (Simmons Royalty was known to be a good employer, and hired both black and white workers.) Upon completion of a well, or when running pipe, additional workers, called "roustabouts," would be called in.[45]

Simmons Royalty was very much the family company, and the father-son team sometimes hired relatives in need of well-paying work (some oil drillers today earn six-figure incomes). Jake Jr.'s brother John, who had helped him two decades earlier in east Texas, worked during the fifties as a "tool dresser." But as he got older the physical strain proved too much, and he decided that, in his words, "I can beat this myself." He returned to Oklahoma City and started a small carpentry business. Jake never stopped admonishing him to strive for more. "Try and do something progressive," he'd lecture his brother. "Be a stepper." To this John would retort, "I'm not a man like you—I'm a lover boy." John measures his success by his relatively sound health and the joy his mellow disposition has brought him. "Wealth don't mean nothing," he says. "I take each day one at a time and by that you get to see more and you hear more. Just think—over ninety, and I've outlived my entire family."

J.J. also employed his younger brother Donald. By the time he was a teenager, Donald, who Jake III recalls was "the best worker I ever had," was already approaching his brother's powerful size, and was capable of adhering to J.J.'s strict work rules. Despite the hard labor, Donald had a job whenever he needed one, which was more than some workers could say. J.J. fired many oilfield workers who didn't "measure up." Measuring up to a man of his stature—he weighed about three hundred pounds at this time—was not always an easy thing to do. Time means money on a drilling operation, which can go on for weeks before reaching the thousand-to-three-thousand-foot depth where oil in eastern Oklahoma is found. It was grueling work.

To this day, J.J. regards himself as a tough but benevolent boss who looks after his employees. Like his father, he calls everybody—including his personal secretary—"Miss," "Mr.," or "Mrs.," and as an Interstate Commerce commissioner he expects to be called "Commissioner Simmons."

Commissioner Simmons is a very serious man—perhaps even more serious than his father. He rarely jokes, and approaches his work with a gravity that borders on severity. To Simmons government service is a weighty responsibility, and authority is not something to be treated lightly.

J.J. III has spent nearly half his professional life in government service. From 1961 to 1970 he held a variety of important oil-related jobs at the U.S. Department of the Interior, including posts as the domestic petroleum production specialist, and the administrator of the Oil Import Administration. From 1970 to 1982, he worked for Amerada Hess, a major multinational oil company, but even there his employment centered on his governmental expertise: he was the vice president of government relations, and a key adviser, confidant, and friend of Leon Hess, one of the ten richest men in the American oil industry. After taking early retirement from Amerada Hess, Simmons returned to government in 1982, where he accepted a two-thirds salary cut to serve as a commissioner of the Interstate Commerce Commission, a federal agency which regulates the trucking and railroad industry. Although a Democrat, Simmons was a fiscal conservative, and his qualifications were impressive. After a year with the commission, President Reagan appointed him to the post of under secretary of the Department of the Interior—the top official after the controversial Secretary of the Interior James Watt (and the second-highest ranking black official in Washington). For many months, while Watt struggled with controversy, Simmons quietly oversaw the day-to-day operations of the Interior Department. When Watt was replaced by President Reagan's close friend William Clark, Simmons returned to the Interstate Commerce Commission, where he developed a reputation for being his own man by responding to political pressure for immediate deregulation in a gradual, case-by-case manner.

As an employee of his father's company, J.J. III also spoke his mind. His father did not agree with everything he heard, and was unyielding in his counterarguments. By 1951 J.J. was ready to quit the Simmons Royalty Company.

The rift between father and son was caused by their different perceptions of J.J.'s role in the company. "I assumed I was a partner from the beginning," he recalls, "although it took three years to emancipate myself from being a son to being a partner."[46]

In retrospect, J.J. reflects that he had done little to demonstrate his right to become a partner, because his father had all the contacts, the reputation, and most of the capital. Time has a way of mellowing youthful disputes with parents: today Jake III has trouble finding any matter in which the test of time has not proven his father right. But tell that to an ambitious son struggling to emerge from the shadows of a powerful father, and a far different voice emerges. At the time, J.J. was investing all of his spare money in the company's drilling operations, and felt that his geological expertise warranted a full partnership. He let his father know that he would not settle for anything less. "I didn't know where I was going or what I was going to do—I had a family and two children already—but I was really going to leave. I didn't seem to be able to penetrate that wall that he had built up between being a son and being a partner."[47]

Neither father nor son was willing to budge, and the argument seemed destined to destroy the partnership. The standoff precipitated one of the few instances in which Eva Simmons interfered with her husband's business affairs.

"My mother was a very strong but sensitive person," explains J.J. "She was very wise, and knew when to interject herself and when not to. She was not at all combative, but this is the one time she disagreed with him—and me, too. She took the initiative, called us both together at their home. She got us both to reason, because he needed me, and I needed him." With Eva's arbitration, the father and son agreed upon a division of labor, authority, and revenues. "Before it was over she told us both to put our arms around each other and never let these kinds of things happen again. And I can't remember that it ever did happen again."[48]

7

THE ONE THAT GOT AWAY

If Jake Simmons hadn't run previously I probably wouldn't have been elected. They felt they had better make some move before Simmons tried again and won.[1]
—Dr. Jesse Chandler, Oklahoma's first African-American elected to public office since statehood

Jake Simmons, Jr.'s, first trip to Africa took him to Liberia, the continent's oldest black republic. It was a fitting destination. The government of Liberia had been founded in 1821 by a group of black settlers from the United States who called themselves the American Colonization Society. The settlers had the strong support of wealthy philanthropists and abolitionists, as well as the U.S. government. Their influential white backers believed that after slavery ended, the alternative to an unwanted multiracial society was sending free blacks back to Africa.

The experiment lasted more than forty years, during which time some fifteen thousand American blacks emigrated to Liberia. The settlers, known as "Americo-Liberians," controlled the newly formed nation from the capital city of Monrovia (named after President James Monroe), while indigenous natives stayed in the hinterlands. Liberia's leading families dressed their daughters like southern belles, lived in replicas of Virginia mansions, and styled their political institutions after America's. To this day Liberia uses U.S. currency as its own, and Michael Jackson is among the country's most popular cultural heroes.[2]

By the mid-twentieth century Liberia had evolved into a single-party democracy that was the closest thing to a colony the U.S. would ever have in Africa. Between 1944 and 1972 it was ruled by William Tubman, a short, stout man with horn-rimmed glasses and an engaging smile. Tubman tolerated little opposition and created a populist model of authoritarianism which many African heads of state would later emulate.

Liberia's president played his American card well, delivering virulent anti-Communist speeches with sufficient fervor to become one of Harry Truman's Third-World sweethearts. Major American corporations began investing there, and the U.S. Army Corps of Engineers built an international airport. In 1952, when Tubman hosted a grand international extravaganza to celebrate his second term as president, Truman sent four distinguished emissaries to attend.[3]

Jake and Eva Simmons also attended the festival—though not as representatives of the U.S. government. They came as guests of Liberian secretary of defense Earnest C. B. Jones, Simmons's Tuskegee classmate. The Simmonses joined hundreds of other international guests for a week of parades, band competitions, and tribal dances. The celebration culminated in a gala ball at the president's executive mansion. As guests danced the French quadrille (once *de rigueur* in old New Orleans) and drank bourbon, Jake used the opportunity to network. Jones introduced him to President Tubman and, equally important, Richard Henries, the president's close friend and speaker of Liberia's House of Representatives.

Simmons dazzled his hosts with starry talk about Liberia's mineral potential. Multinational companies, he explained, were pouring countless millions into oil exploration in North Africa and the Middle East, transforming backwater nations into kingdoms of undreamed-of affluence. The British oil giants had recently begun exploring in Nigeria, less than a thousand miles away. Why should Liberia not benefit from whatever unknown resources the Lord had blessed her with?

Simmons and a group of select foreigners were taken on official tours of Liberia's mining districts. Liberia had only begun to allow its natural resources to be explored. Firestone had a million-acre rubber plantation, and Republic Steel was extracting more than one million tons a year of the world's highest grade of iron ore. Because Liberia was extremely poor, its president thirsted for foreign investment and granted mineral concessions very cheaply. Republic Steel paid the government less than $50,000 a year for its mines, which yielded millions of dollars' worth of premium iron ore. Tubman made it clear that if a foreign company was interested in making money in Liberia, his government, with its "open-door" policy, was interested in talking. To a deal-maker able to bring further mineral exploration into the undeveloped nation Liberia was wide open.[4]

Simmons waltzed right in. He explained to Liberia's leaders that he had connections at the top of some of the biggest oil companies in America. Simmons was willing to pay for exploration options, and assure Liberia of a fair share of the potentially enormous profits. The fact that Simmons had never negotiated an international deal somehow did not get discussed. What mattered was that he had been properly introduced as a successful entrepreneur of high integrity by a top-ranking official, and that he knew the oil business. Also important, perhaps for the first time in the Oklahoman's life, was the fact that he was a Mason.

Simmons, whose father had also been a member of the secret fraternal order, was a thirty-second-degree Mason at the Muskogee lodge of the Prince Hall Masons, the black American Masonic group chartered in the late eighteenth century. In the United States, where white and black Masons had their own lodges, the advantages of black Masonry were limited. In Liberia, it added an important feather to Simmons's cap. Black Masonry was far more important there than any religious or fraternal order; both Tubman and Henries were members of the "Most Worshipful Grand Lodge of Masons of the Republic of Liberia," an order founded by their ancestors in 1867.[5]

The Liberians, Simmons was pleased to find, were open to negotiations. But, like all international traders, he first needed a local power broker to represent his interests. For periodic fees which eventually amounted to some $20,000, he retained Speaker of the House Henries as his attorney—an arrangement akin to hiring Michael Deaver as a lobbyist while he was still on the White House staff. "You don't get something for nothing in Africa, or any foreign country," observes

J.J. Simmons III. "People feel if you're going to get their natural resources you're going to have to pay your way."

Upon returning to Muskogee Simmons set out to make something of his promises. Getting American mineral giants to expand their operations into a tiny, unexplored African nation was not as easy as he had made it sound to the Liberians. The first thing he had to do was let the corporate brass know he had the right connections. At the end of 1952 Earnest Jones arrived from Africa to join him on a tour of the Southwest's oil belt. Groups of important industry executives listened to appeals for investment in Liberia. Simmons and Jones spoke of Firestone's success there, and the stability and success of the country's black government (they also used the opportunity to denounce apartheid in South Africa). Simmons's conclusion was that the "surface and subsurface of Liberia justifies exploration," but there were no takers. Still, the audience listened, and it was the sort of audience Liberians would have had trouble attracting on their own. Simmons's prestige among the Africans as their man in American oil continued to soar.[6]

Simmons's first encounter with African self-government whetted his appetite for politics. He dove into the civil rights movement with renewed vigor and decided to seek public office once again. His strategy was to campaign forcefully in the black community for a local contest in which low voter turnout was expected. In early 1954 eight of Muskogee's sixteen City Council seats were up for grabs, and Simmons was determined to win one. The election promised to be a lackluster race; Mayor Lyman Beard was running unopposed with a slate of eight Independent Party candidates. Most voters punched his party's ticket, and only three of the eight council seats were being contested.

Simmons hoped that if a record number of blacks turned out he could squeak past the all-white Independent Party and slip into city hall. His campaign for a seat from Muskogee's fourth ward excited the city's African-Americans. Tulsa's black newspaper, *The Oklahoma Eagle,* wrote: "A team of city councilmen who were breezing along to an election drew stiff competition from an unexpected source. Many citizens pledged their support to one of Muskogee's most useful and influential citizens. . . . The eyes of Oklahoma and the nation are on Muskogee. . . . Simmons says he is not worried about the outcome of the election; he feels that as intelligent a city as Muskogee (regardless of creed, color, or religion) should have total representation. He asks everybody to please vote."[7]

Simmons was exactly the sort of black candidate the segregationist establishment feared. Institutional racism, which thrived on subtlety, was most vulnerable when openly confronted. The type of person they least wanted at city meetings was one who challenged their systematic bigotry, one who refused to bear his second-class citizenship in silence. The presence of a man like Simmons would be like a searing mirror. The oil man's defiant gaze indicated that there could be no subtlety about treating him as anything but an equal. Once on the inside he threatened to strip their bigotry of its pretense and reveal it as stark, hateful prejudice.

Ruining Simmons's chances of being elected was not a complicated task. Muskogee's at-large electoral system, in which all citizens voted for council representatives for each city district, allowed the white majority a perfectly legal means of keeping African-Americans out of office. The process had succeeded in beating Simmons in the 1946 election, and it would work in 1954—provided enough whites turned out to vote. Simmons was banking on the possibility of Caucasian voter apathy and heavy black support. To thwart him, the Independent Party needed a way to get out the racist vote. The *Muskogee Daily Phoenix*, the city's largest newspaper, was only too willing to help. The *Phoenix* was so racist that it refused to run ads for black businesses. On election day when Simmons opened the front page of his morning paper he read:

> The election seems to have stirred up a great disinterest among White voters, but there was evidence Monday of heavy campaigning among Negro citizens for J. J. Simmons Jr., a Negro who wants to become councilman of Ward No. 4.
>
> In Ward No. 4, Simmons is running against Hayes Holliday and Dr. Joe A. Teaff, both members of the Independent ticket, and the two men with the highest votes will become councilmen.
>
> Voting in the race for Ward No. 4, however, won't be restricted to residents of that particular ward. In city general elections, all

> citizens vote on all races for council—regardless of ward designations. . . .[8]

The warning worked. Voter turnout for the election was low, but not as low as Simmons had hoped. The mayor's party took every seat in the council.

The next day the front page of the *Phoenix* had this to say about Simmons's losing bid:

> . . . In a race which seemed to attract the most voter interest, Jake Simmons Jr., attempting to become Muskogee's first Negro councilman, was soundly trounced—although he did do well in certain Negro precincts.
>
> Simmons, Hayes Holliday, and Dr. Joe A. Teaff were running city-wide in the election for councilman from Ward No. 4 (northwest). The two receiving the highest number of votes would be the winners.
>
> The final votes, however, don't tell the complete story of what happened in this peculiar race for City Council for Ward No. 4.
>
> The Negro candidate carried five precincts, and at least two of them represented a good example of "single-shooting"—a system under which supporters of certain candidates vote only for that candidate, instead of for two as requested on the ballot.
>
> The theory behind this system is obvious. It means that Simmons, for instance, in one of his strong precincts would receive a huge vote, while the other two candidates would draw blanks or near blanks.
>
> That's exactly what happened, particularly in Ward No. 31-A in the Negro section at 1101 North Third. There, the vote was 150 for Simmons, and zero for Holliday and Teaff, although voters were supposed to vote for two of the three candidates.
>
> In precinct No. 15-B, at Douglas School, also in a Negro section, single-shooting also occurred. In that section, Simmons got 105 as against 7 for Teaff and 10 for Holliday. . . . The early turnout of Negro voters was so heavy Tuesday that City Clerk R.L. "Bob" Davis was forced to replenish the supply of ballots at Precincts No. 30, 31-A and 31-B.[9]

Simmons received 906 votes out of 3,124 cast and a plurality in five of Muskogee's forty-one precincts. Had the system allowed district-by-district elections instead of at-large voting, Simmons would have been elected easily to represent the city's black neighborhoods. His ability to convince nearly all black voters in Muskogee to "single-shoot" their ballots bore impressive testimony to his influence. Ironically, the white establishment took offense at this tactic, ignoring the fact that their electoral system guaranteed the exclusion of black candidates with the machine-gun-shooting certainty of its racial majority.[10]

Muskogee did not elect an African-American to city office until 1962. In that year Jesse Chandler, a respected conservative black doctor, was approached by the city's power establishment to run for City Council on their ticket, headed by incumbent mayor Jim Egan. The city had already been desegregated (due largely to Simmons's efforts), and some whites would no longer vote against a candidate because his skin was black. It had become clear to Muskogee's all-white political machine that Simmons or one of his sons might be successful in his next bid for office, unless it placated the black community first. (At his father's urging, J.J. III ran for City Council in 1958 and lost, although 1,362 people voted for him. He won more votes than any other nonadministration candidate, and easily carried the black neighborhoods.)[11] If a black man was inevitably to hold city office, they concluded, why not pick one they could trust?

Dr. Jesse Chandler was their man. Tall, soft-spoken, a World War II veteran and Mason, he had little record of political activism, and was not the sort to upset anyone's apple cart. The mayor's party financed his simple campaign. As a member of the insider's ticket, Chandler won 72 percent of the vote, becoming the first black since statehood to win a political office in Oklahoma.[12]

Today, in his seventies, Dr. Chandler harbors no illusions about why he was chosen to become the city's first black politician. "If Jake Simmons hadn't run previously," he observes, "I probably wouldn't have been elected. They felt they had better make some move before Simmons tried again and won."[13]

The differences between the beliefs and attitudes of the two men are apparent to Chandler. "The blacks who had been fighting all the years before my time for representation had developed and engendered almost a hatred in some people," recalls Chandler. "Simmons knew that the reason he wasn't elected was because he had spoken out so plainly about the racial situation and its inequities. He alienated people with his abrasiveness; in some ways he could have been more effective if he hadn't been like that.

"He was more confrontational than I," Chandler continues. "One of his favorite expressions was 'We demand' something. My figure of speech was 'We want' something. After all, I don't want anyone demanding anything of me." Neither, it appears, did Muskogee's city fathers, which probably explains why Jake Simmons was never elected to political office.[14]

This is not to suggest that Simmons's political career ended in 1954. His influence, in fact, continued to grow. He realized that his odds of effecting change as a power broker were far greater than his chances of ever getting elected to public office. Until the 1950s patronage and payoffs had completely bought off the state's African-American vote. Consequently blacks had no leverage with which to convince Oklahoma's representatives to adopt civil rights measures. Blacks realized that Simmons's unusual degree of independence allowed him to act truly on their behalf. When it came time to vote, thousands throughout the state turned to him for guidance. He forged their support into a powerful voting block, a block whose power he became increasingly adept at harnessing.

Many politicians started listening to Simmons more out of fear than respect. In the end it amounted to the same thing: influence. The Reverend E. W. Dawkins, Simmons's pastor at Muskogee's Ward Chapel A.M.E. Church, remembers that "when political issues would come up the leaders would all cater to Jake because they'd say, 'We have to do something to keep Jake's mouth shut—if we don't do something to recognize Jake he's going to outtalk us.' They'd never take Jake along because they really liked him; they took him along to keep him from barging into their business."

As an orator, Simmons's ability to move a black crowd was unparalleled in Oklahoma history. When he opposed someone on racial issues he vilified his target in no uncertain terms, leading listeners through the reasons for his opposition. When he spoke for a candidate, he explained his support in the context of the racial struggle his followers wanted to know about. Simmons would travel across the state delivering dozens of fiery speeches during a single campaign. Napoleon Davis, a political activist in his seventies who lives in a rural area about six miles east of Muskogee, recalls Simmons venturing out to a small school nearby to appear before tiny audiences. "I was proud that Jake Simmons was there," Davis recalls. "He was my spokesman. He would fight. In open meeting people would sometimes threaten him, and he'd tell them to their face that 'if you messin' with him you messin' with a keg of powder.' In those days you weren't supposed to be like that; you were a sassy Negro or a smart aleck. But I liked that because he was saying things that maybe I wanted to say and maybe I was afraid to say them."

When I interviewed prominent whites in Muskogee, some suggested that if I wanted to hear a less complimentary view of Jake Simmons's political career I should interview Emery Jennings, a retired junior high school principal who opposed Simmons on the speaker's stump during numerous campaigns. Jennings's soft, articulate speeches appealed to listeners on a more sentimental level than Simmons's. Whereas Simmons inspired a crowd and stirred them to action, Jennings was capable of leaving them in tears. Which is precisely the effect he had on me when I interviewed him in his neat Muskogee home in late 1986, shortly before his death at the age of sixty-six. Because those who referred me to Jennings regarded him as Simmons's bitter adversary I was surprised by what he told me. "I loved Jake Simmons," Jennings said, his eyes moist and his voice rich in sentiment. "If I could have had another father it would have been him. He was the most committed and prolific individual I ever knew. The day they buried Jake I knew it was the end of an era."

Jennings called Simmons a "fighter" who was not afraid of leaning against the political wind. As a prime example of this he cites Robert S. Kerr's 1954 senatorial campaign. That year the NAACP, encouraged by the landmark *Brown* v. *Board of Education* ruling, was pressing its branches in southern and southwestern states like Oklahoma to lobby their senators to change their positions on civil rights. For five years southern senators had avoided all civil rights debate by filibustering every time the subject came up. The NAACP campaigned nationwide to get a majority of senators to support "cloture," putting an end to filibustering and forcing the civil rights debate into the halls of Congress.

Senator Kerr voted against cloture four times in 1949–50. (He also voted against integrating the armed forces.) Oklahoma's black leaders pleaded with him to change his stand, but he refused. When it came time for reelection, a handful of important blacks gave their support to Kerr's opponent, Roy Turner, in the all-important July 4 Democratic senatorial primary.[15]

Turner was a former governor and millionaire cattleman from western Oklahoma. He had the support of many ranchers and voters in western Oklahoma and Oklahoma City. Kerr, from eastern Oklahoma, had the oil industry on his side, and many supporters in the eastern part of the state, which included Muskogee and Tulsa. From the outset it promised to be a hotly contested, expensive race, so close that even the black vote would be significant. Jake Simmons, Jr., sided with Roy Turner, and entered the roughest political battle of his career.

Robert Kerr was known as the "Uncrowned King of the Senate." He was the number-two man on the Senate Finance Committee and probably the most powerful national politician ever to come out of Oklahoma. A fiery debater whose oratory was said to be as "poisonous as a scorpion's tail," his opposition to a bill was referred to in the national press as a "kiss of death."

Kerr was an immensely popular man in Oklahoma. He was also the richest man in Washington, with a personal net worth exceeding $50 million. While in office he had no qualms about using his influence to promote the interests of Kerr-McGee, the giant mineral company he headed. With Kerr in the Senate the company increased more than tenfold in size, thanks partly to an unusually profitable $300 million federal uranium contract Kerr brought in.

This only made him more popular in Oklahoma, where he let everyone know that 30 percent of his munificent income went to the Baptist Church. When not in Washington Kerr toured Oklahoma, making twenty-five speaking engagements a month. He was a first-rate entertainer whose intellectual parlance changed considerably once he hit the countryside. Ripping off his jacket and tie, he would grasp his red suspenders while talking himself into a frenzy. Out would come a blue bandanna from the rear pocket. Holding it in front of him, he would wipe the perspiration from his glasses, dab the beads of sweat from his long forehead, then tie the bandanna around his head, like a fired-up Baptist preacher.[16]

Yet beneath this "lover of the simple folk" exterior lay a prominent advocate of white supremacy, a man who promoted his campaign as "100 percent American"—the rallying cry of the Ku Klux Klan.[17] It was this dimension of the famous senator which Jake Simmons squared off against. According to Donald Simmons his father's fierce running battle with the senator began during the late forties, when Kerr invited a group of influential black Democrats to his home to win their support. He let it be known, however, that they would have to sit in the kitchen. "My old man wouldn't go," Donald Simmons recalls. "He said he wouldn't eat in the kitchen."

The senatorial primary race began to heat up in late May, 1954, when *Black Dispatch* editor Roscoe Dunjee published a strongly worded critique of Kerr's voting record. Dunjee had confronted Kerr on his cloture vote. Kerr responded, "If the price of your support is based upon my voting for a cloture rule to halt filibusters in the Senate of the United States, I will have to reject your vote." Turner, meanwhile, promised, "I shall cast my vote for a conservative cloture rule [so] that the will of the people may always be democratically expressed."[18]

Dunjee urged his readers to vote against Kerr. "I do not see how any black man on the outside of an insane asylum can vote for Robert S. Kerr this year," he wrote. "His entire Senate record as well as his gubernatorial background is inimical to the interests of minority groups."[19]

The *Black Dispatch*, a weekly journal containing local, national, and international news, was the most widely read black newspaper in the Southwest. Dunjee himself was among the most respected editors in the history of black publishing. Since starting the paper in 1915 he had become one of racism's most articulate opponents, and served as both president of the state NAACP and a member of the organization's national board for many years. He and Simmons found themselves on the same side of many battles. Dunjee became Simmons's most formidable ally. The two became close friends, a tag team of civil rights activism, with Simmons leading African-Americans in eastern Oklahoma, and Dunjee covering the central capital of the state. The two were in no way the only freedom fighters in Oklahoma, but they were, in their day, the most outspoken—and probably the most effective.[20]

To counter the duo's opposition, Kerr brought Ed Goodwin,

publisher of the Tulsa-based *Oklahoma Eagle,* into his camp. Goodwin was a stubborn, street-tough man who rarely let the facts get in the way of his editorials. Kerr advertised in Goodwin's paper, lent him money, and gave him an important patronage job. Mrs. E. L. Goodwin, the late publisher's wife, still remembers what Walter White, the former national head of the NAACP, told a black audience when he visited Oklahoma. "He said that Kerr should spell his name C-U-R," she recalls. "He said that Kerr didn't like Negroes, he liked a Negro—and he meant my husband."[21]

Through mid-1954 the *Eagle,* more gossipy and far less political than the *Dispatch,* published venomous barbs against Turner's supporters. On June 3, 1954, one of its regular banner headlines read: KERR SUPPORTERS EXPOSE PLOT; ANTI-KERR PAID CLIQUE LEADERS ARE EXPOSED. The article warned, "An abortive attempt by a small clique of hired spokesmen to mislead and confuse the Negro citizens of Oklahoma was revealed this week as a few Negro workers began a campaign of misrepresentation against Senator Robert S. Kerr." As evidence of this "political plot," Goodwin cited Dunjee's editorial "tirade" against Kerr. He added, "Immediately after the publication of Dunjee's edict against Senator Kerr, it was learned that J. D. [*sic*] Simmons of Muskogee and a few other purported leaders had joined the Turner forces and had made plans to send sound trucks and paid agitators into the communities heavily populated by Negroes throughout the state."[22]

Senator Kerr may or may not have liked other Negroes—but he knew he needed their support. He curried favor with other African-American leaders by dispensing generous donations to their favorite causes, like the black Baptist church. He even bankrolled the campaign of a prominent black minister who ran for president of a national Baptist organization.[23]

"Senator Kerr figured he had paid the black people of Oklahoma," remembers Jimmy Stewart, then a national representative of the NAACP. Stewart led an Oklahoma delegation to Washington to lobby for cloture early in Kerr's senatorial tenure. "He said, 'Jimmy, I stand a better chance to become a mother than you have to get cloture passed in the U.S. Senate.' "

To Jake Simmons, Kerr was equally blunt. When he and others appealed to Kerr to support cloture, the senator told him, "I have a great deal of respect for you men, but frankly, any nigra vote I need I'm gonna buy."[24]

Kerr enraged civil rights leaders like Dunjee and Simmons. Dunjee struck back in print while Simmons fought him from the stump. Simmons picketed Kerr across eastern Oklahoma, the senator's area of strength, and denounced him at every barbecue he could talk his way into. By the end of June, with just a few weeks until the July 7 primary, the combatants readied themselves for the home stretch.

In certain ways, Jake Simmons, Jr., and Robert Kerr had a lot in common. Both men had emerged from eastern Oklahoma as successful entrepreneurs and Democratic politicians. Both were powerful orators, articulate, effective, and convincing in politics as well as business. Both had sharp wits, impressive minds, and strong ambitions. Both were fierce, stubborn fighters, but both had a casual, winning way with simple people. Had Robert Kerr been born a black man, he might have acted much like Jake Simmons. Instead he was a U. S. senator, and Simmons was one of the few men ever to stand up to him.

On June 25 Kerr was busily campaigning in eastern Oklahoma. He had already made his pitch to the blacks of Muskogee, making sure that his African-American supporters were present, and spicing his speech with his usual friendly comments to people he knew in the audience. The next night, Kerr attended a fish-fry rally in the nearby small town of Okay, and invited a number of key black supporters from Muskogee to attend. The senator delivered the main address and afterward the audience fell into line for the free food. But the master of ceremonies shouted in the loudspeaker that the food was for whites only, and the handful of blacks attending, including a school principal and a few Baptist ministers, had to get out of the line.

When news of the fish-fry incident reached Jake Simmons the next day he called a political rally at Second and Dennison streets, in the heart of Muskogee's black business district. It was a sunny Saturday, and a large group listened as Simmons stood on a podium and told them what had happened the previous night, blaming "Kerr and his henchmen" for the incident. He related the situation to Kerr's sorry record on civil rights, and gave a persuasive argument for Turner.[25]

"Jake made a terrific impact on that campaign," recalls Emery Jennings. "Jake's speech for Turner made such an impact that Kerr had to come back to Maxie Park the next day and make an unscheduled speech in front of the black community."

The senator was furious. Unable to counter Simmons's accusations, Kerr attacked him personally. Painting a picture of Simmons as a wealthy political mercenary, he told the black audience, "I hear you've got a Negro here who carries water on both sides."[26]

"He just tore into Jake Simmons," Jennings recalls. "But what stopped Jake's political energies from ruining him was that he struck it rich. He was independent of the power structure, and never had to worry about losing his job." Jennings recalls that he also opposed Kerr, but that the superintendent of his school district, a friend of Kerr's, called him into his office early in the campaign and discreetly convinced him that his job security might be jeopardized if he spoke out actively against the senator.

A few days after the fish-fry incident, Goodwin's *Oklahoma Eagle* came out with a special political supplement. It included an open letter that had been sent to Jake Simmons by his self-described "friend" H. James Shooter, a director of a black youth club. The "letter" read:

> Friend Jake:
>
> I am somewhat surprised at you leaving your oil business to go out in the interest of a candidate whom you know and have said that he was no good. You have forgotten that you felt he should have been impeached from the governors office. . . .
>
> I judge you are the big editors last stand. My only regrets are I wish I was getting the money you are getting. I need it, but you don't. I happen to know that you are too good a businessman to leave your own business to attend to someone else's unless you were making as much or more money.
>
> I know that you will say on the 7th of July as you have in the past, "Jim, you were right. I felt all the time Kerr would win but when a man makes it sound good to you and fixes you so you can go—you just can't turn it down."[27]

The special issue also included an editorial, written by the "Kerr-for-Senate Club," which probably meant Ed Goodwin. Goodwin may have been especially determined to smear the Muskogee businessman because of an unprofitable oil investment he once made with Simmons (according to Goodwin's son Jim, "My father invested with him in a dry hole, and felt that Simmons made out but he didn't"). At a time when Senator McCarthy's Red-baiting witch-hunt was in full swing, the *Eagle* editorial struck a particularly malicious note:

> JAKE SIMMONS AND THE CAMPAIGN
>
> One of Jake Simmons' friends said that Jake reminded him of that Civil War general who said, "The battle is won by them that gets there furstest with the mostest." He went on to say that Jake would support a Democrat, a Republican or a Socialist in any election for any office anywhere, whoever made him the best proposition with the "furstest and the mostest." Case after case was cited where this politician from the east side of the state was crossing party lines at will, campaigning for good men and bad men with the same fervor, and apparently generally did not pay too much attention to whether this candidate would benefit the state and the Negro race. This observer summed up his analysis by saying, "A lot of Jake's candidates lost and when one did often the Negro race lost, but no one ever reported Jake losing anything, whoever won. Jake is a promoter. He promotes oil royalties, real estate, politicians and anything of value, but most of all, he promotes Jake." The Kerr-for-Senate Club believes that we Negroes of Oklahoma should not be interested whether one white man or one Negro wins, but that Oklahomans, black red and white should not lose the immense gains that have come to us in the last dozen years during Bob Kerr's tenure as governor and senator. DON'T BE LED AROUND BY THE NOSE BY LOUD AND LOOSE TALK AND CONFUSING ISSUES.[28]

No evidence of material gains made by Simmons or any pro-Turner blacks was ever presented in the *Eagle*. Today Goodwin's son Jim, a prominent attorney in Tulsa and the current publisher of the *Eagle,* believes Simmons opposed Kerr simply out of a longing for racial dignity. "To him there was no compromising against wrong," says Jim Goodwin. "That was the measure of Jake Simmons."

The July 7 primary election was a close one, with Kerr winning 238,543 votes to Turner's 205,241. A handful of other candidates also drew some votes, and because there was no clear majority a runoff election was called. Turner's campaign manager thanked Simmons and a few others for helping his candidate win overwhelmingly among black voters, and asked for their continued support. But the runoff never took place. After spending hundreds of thousands of dollars on the most expensive political campaign Oklahoma had ever witnessed,

Turner ran out of money. Kerr had made the stakes too high. His opponent withdrew from the race.[29]

Simmons pressed on with his domestic oil business and made slow progress on the Liberian deal. Finally, at the beginning of 1956, he got word from President Tubman that Liberia was willing to negotiate with him for an exclusive mineral concession, even though he lacked a major corporate backer. By committing himself independently Simmons would be at greater risk, but he would also be in a better negotiating position when he did find an American multinational to bankroll him. He prepared to return to Africa.

Simmons was a daring man, accustomed to dealing with new people in new territories, but he also knew his limitations. In Oklahoma, he had developed a reputation for having an intimate knowledge of the geology and oil history of a good part of the state. Yet he knew little about international business.

He quickly found someone who did. During the 1930s a friend at Texaco had offered him a job representing the company overseas. Before heading back to Africa Simmons called the man, who had become a vice president at Texaco's New York office, and asked him for a favor. "I don't know anything about foreign concessions," Simmons confided. The oil executive invited him to New York for a crash course.[30]

Simmons was a fast learner. After a week in New York scrutinizing copies of Texaco's international agreements, he was ready to go. (He later told his sons that without Texaco's help he would never have gotten his African deals off the ground.) Accompanying him as he boarded a flight at Kennedy Airport was his son J.J. III. Although the journey over the Azores to Liberia took two days, there was no sense of tedium for the two Oklahomans who sat in first-class, engaged in heated conversation. They were heading into a wide-open region, and there was a lot to talk about.

After thirty years of brokering and dealing in the world of white men, Jake Simmons was going to the one continent in the world where the color of his skin was not a handicap. For many years, long before the term had become popular among his countrymen, he had referred with pride to himself as an "Afro-American."[31] Now he was returning to the land from which his ancestors had been taken centuries before, not as a displaced exile seeking to rediscover a lost heritage, but as an American entrepreneur—the first international black oilman in his nation's history.

Simmons and his son spent a month and a half in Liberia. They slept at Earnest Jones's home for a few nights, and visited President Tubman in his lavishly appointed summer mansion, but generally preferred to stay in hotels. (J.J. says he was initially shocked by the lack of respect the wealthy Africans exhibited in their treatment of servants. "They acted the same way the colonialists did," he recalls. "It just points out how weak we all are as human beings with power.")

The Simmonses, with Richard Henries's help, succeeded in convincing the Liberian government that they were the nation's ticket to the future. Jake left Liberia with a twenty-year lease to exploit all of the country's minerals—oil, gas, coal, iron, and whatever else was below the earth—that had not already been ceded to the handful of other companies operating there. This amounted to an area covering a huge portion of the country, extending over more than eight million acres of land and continuing for miles to the international date line in the Atlantic Ocean. It was one of the largest mineral concessions ever granted to an individual.[32]

In return for the concession Simmons paid $25,000, a considerable sum for the tiny government. As with all of its new mineral concessions, Liberia was to receive a 50-percent share of the net profit which the Simmons mining concern earned. Simmons agreed to pay an additional $100,000 annually in lease payments for the duration of his contract. By the time he left Liberia, his expenditures on the deal approached $100,000.[33]

If Simmons was biting off more than he could chew he never showed it. Immediately upon returning to Oklahoma he began finding backers to carry out the mineral exploration of the new territory. J.J. was dispatched to libraries around the country where he studied European geological periodicals for indications of what might lie below Liberia's surface. (He concluded that commercial quantities of oil would not be found there, but that the country was filled with other valuable minerals.)

Simmons covered some of his financial exposure by selling small pieces of his position to a few Oklahoma investors, although he maintained a majority of both the equity and the money at risk. He then set out to find a large oil company with the muscle to invest the millions, or tens of millions, it would cost to explore Liberia. For a 5-percent overriding interest in the deal's profits, he was willing to turn over the concession to one of the big boys.[34]

Simmons found an interested ear in the chief executive's office at the Sunray Midcontinent Company, a Tulsa-based oil company with $330 million in sales. C. H. Wright, Sunray's chairman, and a number of other top executives were familiar with Simmons from local oil purchases they had made from him. J.J. III recalls that he and his father had numerous lunch meetings with Wright and H. O. Harder, Sunray's exploration chief, in both the corporate dining room and the executives' offices. Black businessmen were an unusual sight in such corridors of power. One former Sunray executive remembers "what a shock value it had on the executive floor to walk down the hall and see this black man in the corner office, well-dressed and talking deals."[35]

Jake Simmons, the top brass at Sunray realized, could deliver the goods, but his company was new to international exploration. The execs wanted another local giant, one with far greater experience in mineral exploration, to hold their hand on the Liberian venture. They wanted Kerr-McGee.

Wright was friendly with Senator Robert Kerr and his partner Dean McGee, and was involved with their company on other projects. Kerr-McGee, the nation's largest uranium miner, had organized the best team of heavy-mineral geologists in America. They also had some experience internationally, developing small deals in Cuba and Kuwait. "Bring Kerr-McGee into the picture," the Sunray executives told Simmons, "and you've got a deal." Simmons had never expected to find himself in a position where he needed to talk to Kerr again, much less solicit his support for a crucial business venture. Yet Sunray insisted that there would be no deal without Kerr-McGee. The Liberian government had allowed Simmons a few extensions of his annual $100,000 lease payment, but time was running out. After his initial outlay of cash, he was in no position to risk much more. He needed a corporation's deep pocket to foot the bill. Simmons knew that if he could not win Kerr-McGee's backing he would have to abandon his enormous international concession.

According to J.J. III, who was present at the negotiations, a meeting took place at Sunray's headquarters in which Kerr left the conference room to huddle with Sunray's top executives. When they came out, the deal was dead. J.J. believes that "because of the political wars which my father and Senator Kerr had fought, Kerr killed a deal that Sunray desperately wanted to go ahead with."[36]

Unfortunately neither Senator Kerr nor any of the Sunray executives present at that meeting are alive today. Through a company spokeswoman Dean McGee, the aging cofounder of Kerr-McGee, took exception to J.J.'s recollection. According to McGee, "The decision not to go ahead was an economic one. We were extended as far as we could in the Gulf of Mexico and expanding our uranium exploration, and couldn't participate because we didn't have the money and have a conservative policy in regard to spending."[37]

Sunray later built a $12 million refinery in Liberia. It was one of the company's first projects abroad, but the man who introduced them to Africa had no part in it. It was in some way appropriate that Jake Simmons should fall hard the first time out in the tough world of international deal-making. The qualities which allowed him to excel were not blind luck and opportune positioning, but persistence and determination.[38]

"He never got discouraged about anything," explains J.J. III. "He was just an eternal optimist. He would drill quite a few dry holes in Muskogee County. I'm sure it must have bothered him, but he felt the next well was going to be a thousand-barrel [-a-day] producer. And it's the lessons he learned in Liberia that served him well in acquiring concessions in Nigeria."

It was an expensive lesson—more expensive in what Simmons didn't make than in what he lost. Hundreds of millions of tons of iron ore, worth billions of dollars, have since been discovered in Liberia by various mineral companies. A considerable quantity of diamonds are also being mined. Simmons could have had 5 percent of much of this if his partners had gone through with the deal. As for oil, no commercial quantities have ever been found in Liberia, either on or off shore.[39]

8

BACK TO AFRICA

Jake recognized that the people who were going to put big dollars into big oil were not going to be looking for forty-acre prospects in Creek County, Oklahoma. They were going to be looking for billion-barrel oilfields on the continental shelves of the world. The trick was to get himself situated so he could get between the reserves and the money. And that's exactly what he did.[1]

—W. H. Thompson, former president, Signal Oil & Gas Company

The downtown skyscrapers of Bartlesville, Oklahoma, rise like Oz from the endless miles of flat countryside surrounding it. Located forty-five miles north of Tulsa, the city is a crossroads to nowhere. Freight bound for points beyond rarely passes through the community of 37,000, and few tourists visit its fine western museum, which houses the world's largest collection of Colt firearms. Because the name of Phillips Petroleum, America's ninth-largest oil company, has far wider recognition than the name of the city in which it operates, Bartlesville's Chamber of Commerce promotes itself with the slogan "Known by the company we keep."[2]

Phillips Petroleum also bears the distinction of being the first company ever to "keep" Jake Simmons—at least as much as any company can be said to have kept a man who refused, through his entire professional life, to accept full-time employment from anyone. Perhaps Simmons's mutually beneficial association with Phillips can best be described as a partnership—a partnership between one of Oklahoma's most unusual entrepreneurs and the largest company ever to rise from its prairies.

Bartlesville epitomizes the modern company town more than any place in America. Phillips Petroleum directly employs more than a third of the city's workers. In return for a commitment which encompasses an employee's social, religious, and community life, the company provides everything it believes a middle-American family needs for its happiness. A vast labyrinth of recreational facilities lies below Phillips's downtown buildings, containing restaurants, cafeterias, sports facilities, a billiards room, a card room, an Olympic-sized pool, and even a quiet room with well-cushioned chairs and darkened lights for off-duty naps.

Bartlesville's babies are born in the $45-million Jane Phillips Medical Center. When they retire citizens can look forward to attending symphonies conducted by a professional fifty-person Phillips-supported orchestra in the state-of-the-art community center that the company helped build. To Phillips employees and their families, explains one Bartlesville native, allegiance to the company comes first, with America a close second and the state of Oklahoma a distant third.

In a sense Phillips Petroleum is to Bartlesville what the Mormon Church is to Utah. Like the Mormons, Phillips sends legions of wholesome, fair-skinned corporate missionaries around the world. Although they spread the "good news" of the company across the three dozen countries in which Phillips operates, few of their dark-skinned customers or associates find their way back to the headquarters in Bartlesville. For the most part white faces conform to the white shirts worn by the company's formally attired executives. Phillips is so much the paradigm of white-bread middle America that as recently as mid-1986, when I lunched in the company's executive dining room, every one of the 150 or so darkly suited executives present was not only white but male as well.[3]

In 1964, when Jake Simmons was first commissioned to represent Phillips Petroleum, civil rights and integration were alien notions in Bartlesville. Its hundreds of black citizens lived in "nigger town," a small district between Bartlesville's downtown skyscrapers and the foul-smelling zinc smelters to the west of town. When a local minister came out for integrated housing, members of the generally peaceful community barraged him with death threats. Sally Dennison, the forty-year-old chairman of Council Oak Books (a Tulsa publisher) who grew up in Bartlesville, recalls that her family's housekeeper always went to Tulsa on her days off because there wasn't a single social establishment in her hometown that admitted blacks. "It was like they wanted the blacks to just vanish, to evaporate when they were through cleaning up the house," Dennison explains. "They wanted them to be totally invisible."[4]

Jake Simmons's first trip to Bartlesville had little to do with civil rights. Establishing a business relationship with Phillips Petroleum, the largest energy company in Oklahoma, was an alluring objective to any independent oilman in the Southwest. His early business contacts with Phillips were at a relatively low level, selling small amounts of crude or perhaps local leases, to the company's domestic department. He was also acquainted with some of the company's top executives, including officers in the international exploration division.

Like ambitious oil brokers everywhere Simmons wanted to assist in matching a major oil company with a concession from a foreign government. Independent oil traders the world over dreamed of duplicating the phenomenal success of Calouste Gulbenkian, the Armenian deal-maker extraordinaire. Gulbenkian earned an incalculable fortune, and the nickname "Mr. Five Percent," during the 1920s by tying up a good part of the productive acreage in the Middle East, then handing it over to the major multinationals in return for a 5-percent override on all the oil they produced there.[5]

Simmons's first international offer to Phillips sounded much like the solicitations made by countless other peddlers. In early 1963 he wrote to the international exploration chief, offering his assistance to Phillips in negotiating a concession it was working on in Libya. He also proposed to put the company in touch with an influential Syrian businessman who could pull strings behind the scenes, although the help would not come cheap: the Syrian wanted a $10 million fee for his services. International production manager Owen Thomas responded that his company had their own personnel, and a significant amount of land already secured from the Libyans. In that case, Simmons persisted, did they need his help in Liberia or Nigeria? "We'll keep you in mind," Thomas replied coolly.[6]

Thomas recalls that the impression Simmons left was "outgoing, energetic, and likable—he presented himself in a very forceful way." On the first go-round, however, Simmons's character alone was not enough to win Phillips's business. "We had all kinds of offers like that over the years," explains Thomas. "We turned almost all of them down."

By 1963 Jake Simmons was a man to be turned down politely. His political influence had grown with the civil rights movement. As civil rights became a mainstream issue Simmons became a mainstream leader, and his family became well known in the Southwest. After John F. Kennedy won the Democratic nomination in 1960, his brother Robert called a group of black leaders to St. Louis to win their support against the Republican nominee, Richard Nixon. Both Jake Simmons, Jr., and his son J.J. were among those who attended. Frank Reeves, the campaign's minority-affairs liaison, had told Robert Kennedy about the Simmons family's reputation. At three in the morning, J.J. heard a knock at the door of his hotel room, followed by "I'm Robert Kennedy. I promised I'd come visit you. I'd like to come in and talk." The two men met in Simmons's room. Kennedy explained that his brother's administration wanted to integrate federal agencies. "I've never met another black person who was a geological engineer," Kennedy remarked. "We're going to win and would like you to be a part of the Kennedy administration."

After the election Robert sent J.J. an invitation to his brother's inauguration. They met for breakfast and Kennedy offered him a post in the Department of the Interior. In 1961 J.J. left Simmons Royalty in his father's hands to become a "regional oil and gas mobilization specialist" in Battle Creek, Michigan.[7]

J.J. stayed in Battle Creek for fifteen months, living down the street from his sister Blanche. He was then promoted to domestic petroleum production specialist for the Department of the Interior in Washington. He would never return to his father's business. Jake soon began scaling down his domestic drilling activities and pursuing other opportunities in the oil industry. He never asked his son to come back to Muskogee, and voiced approval of J.J's career move. Privately, however, he confided to Eva that J.J.'s departure depressed him terribly. "My mother told me this much later," J.J. explains. "I didn't have the slightest idea at the time because he always expressed so much pride in the fact that I was coming to government. He was a man who had a lot of vision and understanding in that he wanted to support a person in whatever he did."

For the Simmonses success was like a rising tide. As Jake's sons achieved positions of influence they became empowered to use that influence on their father's behalf. The father, in turn, used his new opportunities to promote his sons. On January 1, 1963, Senator Kerr died unexpectedly, leaving behind an open seat in Congress. Governor Howard Edmondson, Ed Edmondson's younger brother, resigned his position and was appointed to take over Kerr's Senate seat. For the first time in nearly a century African-American staffers were being listened to in Washington, and Simmons had little trouble convincing Edmondson, a family ally, to hire his son Donald. At twenty-eight Donald became a legislative assistant—one of the few blacks in the capital to hold such a job.

In many respects Donald was the most headstrong of Jake Simmons's three sons. He was strong-willed and independent, which often led to rebellion against his solemn father. Working on Edmondson's staff helped define Donald's career path. It also taught him the art of political diplomacy, a skill which eventually served both him and his father well in Africa.[8]

By helping J.J. and Donald become acquainted in Washington, Jake Simmons became better plugged in himself. When it came time for the federal government to send its first official trade mission to East Africa, Jake Simmons's name, previously unknown in the nation's capital, was brought to the attention of the trade mission selection committee.

Roy Gootenberg, a former instructor at Harvard University's School of Government, was appointed by President Kennedy to strengthen the Department of Commerce's international trade missions in late 1961. Before long dozens of missions, each containing five to seven successful American businessmen, were being sent to the far corners of the earth to promote American enterprise and products abroad. Thousands of businessmen competed for the few openings as "mission members," an honor which came complete with an all-expense-paid, three-to-six-week journey—and the opportunity to make important contacts.

Gootenberg does not remember exactly who recommended Simmons to the selection committee, which solicited congressmen, governors, and various organizations for candidates. He thinks it likely that one of Simmons's political allies, perhaps Ed or Howard Edmondson, nominated Simmons for the U.S. trade mission to East Africa in late 1963. The mission was one of particular diplomatic importance, since the four nations to be visited—Uganda, Kenya, Tanganyika, and Zanzibar—had only recently emerged from colonial domination. Maurice Keville, then a staff assistant to President Kennedy, remembers approving Jake Simmons while "purifying" the list of potential mission members for the White House. Keville recalls that Jake Simmons carried considerable influence in the Kennedy administration. He passed the list on to Gootenberg at the Department of Commerce for the final cut.

When Gootenberg examined Simmons's credentials, he was overjoyed. At the time his department had been able to place only one other black entrepreneur on a trade mission. "Blacks usually weren't in the type of businesses that we were interested in selling abroad," the former director of international trade missions says. "Most of their companies were small and might have been successful, but only in service industries within the black community, like undertakers and insurance companies. The Africans often felt patronized or condescended to. They didn't want to see black businessmen [just because they were black]. They wanted to see businessmen of some influence in the United States. So Jake was a find: he'd been successful in an area where white business dominated and white business was not having anything to do with black businessmen."[9]

The international trade mission was an opportunity for Simmons to be an official representative of the U.S. government in Africa, and he was determined to milk the trip for all it was worth. He saw it as a springboard to renewed involvement in Africa. As Eva helped him prepare for the trip, he said, "I am going to Africa to get me an oil concession."[10]

The time had come for Simmons to pull a few strings. He knew that the East African nations he was scheduled to visit were not where the oil action was. Fantastic new discoveries of top-quality "sweet crude" (high gravity and low in sulphur) were being made every month in Nigeria, but a trade mission had already visited there. Simmons decided not to let this stop him. There was nothing to prevent him from going to Nigeria as a private citizen—though a well-connected one—after his official business was done. He visited G. Mennen Williams, a politically liberal friend in Washington who had been appointed assistant secretary of state for African affairs. Would the assistant secretary mind, Simmons asked, assisting an American businessman abroad by sending a cable to the U.S. ambassador in Lagos, Nigeria, notifying him of the Oklahoman's imminent visit and intention to secure a new oil concession?

Williams was only too happy to comply. Next Simmons used the trade-mission appointment as an excuse to make another pass at Phillips Petroleum's international division. Three weeks before the mission was to leave, he contacted Phillips's New York office, telling them that after his trade mission to East Africa he would be free to travel on Phillips's behalf—provided the company absorbed his expenses in Nigeria.

Now Phillips executives were more tentative in their refusal. "At this time we do not believe we have need of your services over there following your trip," the New York representative notified him. "However, we appreciate very much your keeping us in mind as it is very possible that sometime in the future we might be able to work out something to our mutual advantage."[11]

It was not exactly a commission to proceed, but neither was it a definite no. Jake Simmons construed the response as an almost yes. With a little polish and embellishment, he had himself a backer.

The East African trade mission left Washington on October 26, 1963. To the group's African hosts Simmons was the center of attention. He met with the heads of state of each nation visited, and was granted a private audience with Kenya's powerful president, Jomo Kenyatta. In Dar es Salaam, the capital of Tanganyika, Simmons was singled out by the government to address a special assembly of Tanganyikans interested in the role of blacks in American society. Unlike Malcolm X, who visited Africa a few months later and immediately denounced America's government as being worse than South Africa's, Simmons came off as a patriot. While a U.S. representative, he suspended his opposition to racism at home and provided a glowing report on the progress of blacks in the United States. His speech focused on the opportunities capitalist America offered its citizens, both black and white. It could just as easily have been written by Booker T. Washington.[12]

For the U.S. government, according to trade commission chief Roy Gootenberg, Simmons's participation in the mission "had a real public-relations impact in Africa." The Voice of America sent a reporter to Dar es Salaam to produce a special segment on his speech, which was transmitted all across the African continent on December 4, 1963. The broadcast opened with "An American Negro businessman has told an African audience that Americans of African descent have had conspicuous success in American economic life." One excellent example of this was Simmons himself: "Mr. Simmons's business is mostly with white American customers, and he cited this to show that Negroes are in business in America not only to serve their own people . . . but to compete for business in all sectors of American life."[13]

Ambassador Horace G. Dawson still recalls Simmons's arrival in Uganda as part of the trade mission. Dawson, then cultural attaché at the U.S. Embassy in Kampala, was among the first black American officials to be posted in Africa. He and his wife met Simmons at a special reception given by the American ambassador. "We were fascinated by him because he was such an energetic person and a nonstop talker," Dawson remembers. "He was very interested in meeting the Africans and not just with the idea of selling. He was always a man with a mission about what black folks should stand for and do. He talked to us and the African businessmen he met about his own background and how he became successful. We didn't know anybody black who was successful in the oil business or anything about blacks in Oklahoma and the whole Indian background. All that was new to us."

Simmons's ability to charm and fascinate East African dignitaries earned much goodwill for the trade mission but did little for him personally. When the assassination of President John F. Kennedy on November 22, 1963 caused the East African trade mission to disband one week ahead of schedule, Simmons thought about returning to the states with the others for the funeral—then thought again. He instead hopped a plane for Lagos, Nigeria, and telephoned the American ambassador for an appointment.

The telegram from his ally Williams in Washington won Simmons quick access to the U.S. Embassy. He asked the ambassador to follow the State Department's instructions and arrange an appointment with Yusuff Maitama-Sule, Nigeria's influential minister of mines and power, to discuss potential oil concessions for Phillips Petroleum.

"I'm afraid you made the trip for nothing," the ambassador advised. "There are no oil concessions available in Nigeria. The major oil companies have them all."[14]

According to Donald Simmons, his father asked the ambassador if he knew anything about the oil business. The diplomat admitted he did not. "Well, if you did," insisted Simmons, "you know every major oilfield in America wasn't found by majors but by independents like me. Let's make a deal: I won't pretend to know anything about the diplomatic business and you don't presume to know anything about the oil business."

The ambassador, though peeved, arranged the appointment. He accompanied Simmons to Maitama-Sule's modern office and had second thoughts about his initial pessimism when he saw how the minister, clad in African national dress, greeted the Oklahoman. "I didn't know an African-American was in the petroleum business," remarked Sule, rising from behind his desk and putting his arms around his unusual visitor. "You are most welcome to my country, my brother." Simmons and Sule continued their meeting in private throughout the day. Simmons told the minister that he had Phillips Petroleum behind him, a company with plenty of cash to carry out the expensive exploration requirements of a Nigerian concession. Sule was duly moved, both by the Oklahoman himself and the "company he kept." In an interview a few years ago, Maitama-Sule said that Simmons's outstanding reputation, which he had checked out by his staff, persuaded him to do business with the first American oilman of African descent that he had ever met. "I told him there was not anything immediately, but as soon as there was something I would be quite willing to consider it," recalled Sule.[15]

Roy Gootenberg still talks admirably of Simmons's diplomatic savvy. "He sort of bluffed his way that he was a representative of Phillips Petroleum," Gootenberg says. "Here was a swashbuckling independent black oil operator who writ large by becoming a member of an official U.S. mission, then going to Nigeria with his credentials and getting in on the ground floor of an exploration franchise [by] representing an oil company that had not had him on their rolls at all." But Simmons knew that he had scored a prospect which few partners could refuse.

* * *

When Jake Simmons returned to Bartlesville, executives at Phillips Petroleum viewed him in a whole new light. In the international end of the oil industry a bird in the hand is worth twenty in the bush. To an American company looking for a piece of the Nigerian action Simmons's promise from the minister of mines was an especially rare bird. Owen Thomas, Phillips's former director of foreign exploration, remembers, "He had something. In Nigeria he actually had a concession. That made a difference."

It was an important enough difference to break the color barrier. Just barely. The authorization for allowing Phillips to become the first major oil company in the world to be represented abroad by a black American had to come from the top. It took incessant lobbying at the Bartlesville headquarters to convince the skeptics to commission him. At the forefront of the argument was the man who became one of his most important allies, Edwin Van den Bark.

Ed Van den Bark, who had just been appointed head of all international operations at Phillips, was not your run-of-the-mill paper-pushing executive. Born dirt-poor in rural Nebraska in 1917, he earned his geology degree at the University of Nebraska through the ROTC and went to work for Phillips in 1939. He rose to head domestic operations, then was transferred to international. Under his stewardship the international exploration wing mushroomed in size, a fact which largely explains Phillips's rapid growth from a $1.4 billion company in 1964 to a $16 billion company in 1982, when "Van" retired. He was credited with increasingly significant finds in Egypt, Libya, Nigeria, and eventually, the North Sea, where he spearheaded an enormously risky exploration effort resulting in one of the largest strikes in history.[16]

Van den Bark discussed Simmons's career with me in his impressive home, exquisitely decorated with Third-World art which he and his wife collect on their constant trips abroad. Plain-spoken and unpretentious, Van den Bark never became a director at Phillips. He was too independent for that, too little the company man. What's more, he was an actively progressive Democrat in an organization of staunch Republicans. Ironically it was Van's unpopular liberalism which contributed to the company's international success. "They said, 'Here's the international department, take it and run,' " recalls Van den Bark. "So I had to make my own rules. There was concern at the company: How do you treat an Algerian? How do you treat somebody from Japan? How do you treat a Nigerian? I decided to treat everybody alike and that proved to be the right answer because we were very successful in getting acquainted around the world fast."[17]

Van den Bark's lack of racial prejudice made him a fair boss but he didn't hire Simmons out of a desire to integrate the international department. He hired Simmons partly because of his prospective concession, and also because Phillips was in a crunch, and he realized that the Muskogee oil operator might provide him with a ticket out of it.

Although Phillips executives were reluctant to reveal the details of the Nigerian deal, when I combined their responses with the company's records and Donald Simmons's recollection of what his father told him, it was possible to piece together a rough picture of Jake Simmons's first successful African negotiations.*

The first companies to search for oil in Nigeria were British Petroleum (BP) and Shell (an English-Dutch amalgamation). As British colonial subjects the Nigerians had little choice in the matter. Immediately after achieving independence in 1960 Nigeria's officials set out to reassign the British oil monopoly's colonial-era mineral concessions.

One of the first companies they turned to was Italy's national oil company, known as AGIP. Enrico Mattei, AGIP's suave chairman, had built a reputation as the "bad boy" of the industry, an anticolonial David who sided with oil-producing nations against the oppressive Goliaths he called "*le sette sorelle*."[18] He popularized this "Seven Sisters" characterization of a nefarious world oil cartel (consisting of Exxon, Shell, British Petroleum, Mobil, Texaco, Gulf, and Standard Oil of California) and advised oil-exporting countries to shop around for a noncartel deal. AGIP, Mattei argued, was that better deal.

The Nigerian government agreed. In 1962 AGIP was promised a former Shell concession of more than a million acres in the heart of the oil-rich Niger delta.[19] But AGIP was much better coming up with nationalistic rhetoric than cold cash. In Nigeria, where wells ran twelve thousand feet deep, they found themselves in over their head.

* As with all the passages in this book dealing with business in Africa, some of the figures used are my own estimates.

After spending more than $50 million on two dry holes, AGIP officials decided they needed a partner. A deal was soon struck with Phillips Petroleum. In return for a passive 50-percent interest in the stock of Nigerian AGIP, Phillips would pay half the cost of the two dry holes and contribute 50 percent to the cost of all future wells. AGIP, which maintained a field office in Nigeria and employed local workers for its rigs, would continue on as well operator, and receive a 12.5 percent stake in Phillips's unexplored North Sea concession.[20]

Everything went fine for the new partnership until oil was discovered. The first well which the AGIP-Phillips team drilled struck oil—big oil. It was one of the most productive wells in Nigeria, and it came with the right to drill dozens of additional wells nearby. Suddenly what had been regarded by Phillips as a passive investment demanded much more active participation.

The company became worried. Phillips executives never expected such a quick success. Their deal with AGIP had not yet been authorized by the Nigerian government. They were beholden to AGIP to honor its agreement, and when billions of dollars are at stake, honor becomes an unreliable quality. Furthermore, AGIP regarded Phillips's participation as a "stock" interest. For Phillips's accounting purposes, unless it got Nigeria to grant it a "working interest" it would eventually cost the company as much as $15 million a year in U.S. taxes.[21] But the Bartlesville company did not have an office or regular representative in Nigeria. Their only contact was through AGIP. And AGIP, all of a sudden, was being less than cooperative.

How uncooperative they were is hard to say. L. M. Rickards, current Phillips executive vice president, plays down any past difficulties with AGIP. Given the company's continuing involvement with AGIP in Nigeria, it is unlikely he would publicly state otherwise. But Owen Thomas, the retired head of exploration, helped negotiate the deal. He says that his department was scared that Phillips might not even have ended up with a stock interest. "We suspected that AGIP was trying to stop us from getting it approved," he explains. "We never let them know we were suspicious, but internally we had some fears." Donald Simmons is perhaps the most candid in his analysis of AGIP's position. "They tried to scuttle Phillips in Nigeria because they wanted to have the whole thing for themselves," Simmons says. "The Italians didn't give a shit—if they had to bribe a minister they would have no problem with that, they had no problem getting done what they needed to get done. We put people in jail for that here; in Europe, if it's in the national interest, they get commended for it."[22]

The AGIP problem was surfacing just as Simmons was visiting Bartlesville to pitch his own potential deal in Nigeria. He could not have arrived at a better time. Nigeria was seen by Phillips officers as a crucial step in what *Business Week* called an emergence "from its provincial shell to become a full-fledged international operator."[23] Van den Bark realized that Simmons would be the perfect person to assist in the crucial Nigerian AGIP negotiations—and press for a new concession that Phillips could operate on its own.

The only blacks working for Phillips in 1964 were waiters in the company's guest suites. Van den Bark countered a barrage of snide remarks about hiring Simmons by telling his associates to "just shove it." He attributed their opposition to what he calls "race jealousy," and managed to prevail. Logic was on his side. The company's provinciality worked against it in Nigeria. Phillips suddenly had to take a careful drive into uncharted territory with blind men at the wheel. "We didn't have anybody experienced in Africa," reflects Van den Bark. "We were young and neophyte in international operations and knew Jake had experience in Africa from his work with the government on the trade deals, so we selected him to assist us in getting our permit from the Nigerians approved."[24]

To the credit of Phillips, Van's rationale overshadowed any racist opposition from the staff. Simmons did a lot of the convincing himself. To those in power his contacts from the trade mission made him more than just an oil industry novelty. Before the trip, he told an associate, he had trouble getting in on the ground floor at Phillips, but afterward he "got right up to the executive suite." Once he got his foot in the door he swung it wide open. Simmons became a familiar sight on the company's penthouse floor, coming and going, as one executive noted, "like he owned the place."[25]

Simmons managed to find friends in all the right places. His most powerful booster was William W. Keeler, the fifty-five-year-old chairman of Phillips's powerful executive committee. Keeler was intensely proud of his part-Cherokee background, and served as principal chief of the tribe for many decades. He loved talking to Simmons about Indian history and his Creek heritage. Executive Vice

President Len Rickards still remembers how Keeler first introduced him to the Muskogee oilman in 1964. The boss called him and an associate into his office to meet "the most prominent member" of the Creek tribe. "We walked in and here was this very distinguished looking black man," recalls Rickards. "And I sort of looked around as if to say, Where is the Indian? Bill told us that one of Jake's relatives had been chief of the Creek tribe."[26]

Once Simmons's role was approved he met with Van den Bark and his staff to discuss the AGIP deal and the possibility of new concessions. The international department had decided that the time had come to open a permanent office in the Nigerian capital of Lagos, and Jake was to be dispatched as corporate ambassador. "We coached him to talk Phillips," Van den Bark recalls. "We wanted Phillips to be known and he was a good man to spread the gospel."

Simmons returned to Lagos in the spring of 1964 and went straight to the office of his friend Yusuff Maitama-Sule, the minister of land and mines. Sule probably never realized that Simmons had been shooting from the hip five months earlier when he claimed to be representing Phillips Petroleum. Now he had returned, as promised, with the backing of an oil multinational and Sule, true to his word, had laid aside hundreds of thousands of acres for a new concession.

Simmons became Phillips's advance man, mapping out contacts and territory so the company's executives would be able to find their way. He had little trouble adapting. In a land where ancestors are worshiped, his love of relating his own family's history charmed his hosts. In a culture where storytelling is an honored craft, his oratorical powers impressed them even more. He dressed well (in conservative western attire), ate everything put in front of him, and took an earnest interest in his friends' families (an essential part of African protocol). He was everything his Nigerian contacts wanted an African-American to be. What's more, he was a black nationalist.[27]

Ambassador Horace Dawson remembers Simmons's pet complaint about Africa during the sixties. "He couldn't quite understand why, if the Africans were independent, they didn't take over the reins and run everything," recalls Dawson. "In those early days of independence Africans were relying upon the British, the French, and all these expatriates to help them run things. Simmons thought they ought to be totally in charge, with fewer whites running vital industries and government. He resented having to deal with those white bureaucrats in Nigeria to get oil concessions. And he didn't like the idea of coming into a country and finding a white immigration official."

Simmons never identified with Marcus Garvey's back-to-Africa movement. He was not a separatist; he wanted to see blacks participate fully in the American political and economic system which he patriotically espoused, not escape from it to other countries or detached subcultures. Malcolm X, the Black Muslim leader whose radicalism had, by the early sixties, grown to challenge the traditional civil rights movement, believed in changing the system completely. In a speech at a Harlem rally in 1964, Malcolm X gave his view of capitalism: "The capitalistic system cannot produce freedom for the black man. Slavery produced this system and this system can only produce slavery." To Malcolm X, white American businessmen in Africa were working to export capitalistic subjugation. "It had been her human wealth the last time," he noted. "Now [they] wanted Africa's mineral wealth."[28]

Jake Simmons, on the other hand, was what his son Donald calls "a complete 100 percent capitalist who made no apologies about it." Nonetheless, when Jake Simmons met America's best-known black nationalist in the lobby of Lagos's Federal Palace hotel, the two men found a lot to agree on.

Malcolm X visited Nigeria for a few days after making his first pilgrimage to Mecca in April 1964. He was treated as an important dignitary there, and made appearances at Ibadan University, on radio, and on television. There is no way to know how Malcolm X viewed Jake Simmons's unusual role in international commerce. The only recollection of their meeting is what Donald Simmons was told by his father. "He talked with him for most of an afternoon," Donald recalls. "He said he was extremely impressed by his intelligence and commitment to black people."[29]

Although he was moved by the conviction of Black Muslims, Simmons—and many Nigerians—disagreed with Malcolm X's hatred of the international exploiters of Africa's minerals. Government officials viewed competitive development of their oil resources as Nigeria's greatest hope for economic prosperity. Local investment capital was scarce, and companies like Phillips were willing to invest hundreds of millions of dollars while cutting the Nigerian government in for half their profits. Maximum participation with minimum financial risk was the objective, and the oil ministry came close to realizing it: by 1977, Nigeria was earning about $20 billion annually from its oil resources, accounting for 82 percent of all government revenue.

Phillips's interest was overshadowed by far larger and more experienced companies like Gulf, Mobil, British Petroleum, and Shell, not to mention better-connected operators like AGIP. It was Simmons's job to persuade the Lagos authorities that the Oklahoma company merited a spot on their busy dance card. There was nothing truly extraordinary to recommend an international lightweight like Phillips to the Nigerians. Nothing, that is, except Jake Simmons.[30] When I interviewed oil minister Maitama-Sule by telephone in August 1982 he was Nigeria's ambassador to the United Nations. What had impressed him about Simmons, he recalled, was the Oklahoman's record. "In Nigeria we could not afford to allow adventurers to engage in the oil business," he explained. "I was convinced that Mr. Simmons was not an adventurer. He had worked his way up the hard way and had established an excellent reputation. He was an honest businessman determined to do an honest job and to get his honest profit. I was impressed by this and the fact that as a black man he was in the oil business. That is why I was most encouraged to give him my full support."

Owen Thomas, who traveled to Africa with Simmons on Phillips business half a dozen times during the sixties, explains Simmons's role: "We met the top government official, Maitama-Sule, through Jake. He spent a lot of time in West Africa and, like a politician, was astute in finding out who to know. Being black he hit it off with them and there was a certain trust there that they might not have had with us as white people from a foreign country. He introduced us to all the key people in a very short time and helped us expedite matters."

Ed Van den Bark recalls that Simmons did more than make introductions for the AGIP deal. "Most of Jake's time," says Van den Bark, "was spent making acquaintances for us in the government. The other part of the time was spent with my exploration manager [Owen Thomas] explaining to the Nigerians all the details of our agreement. And that required lots of work because the Nigerians weren't experienced at all. They had only been independent for about five years then, and had arrived in a position in the Third World where they were trying to learn as much as they could very fast. Looking back on it now I think they did an excellent job, but it took a lot of patience explaining to them how we were forming this company and how stock originates and what it meant to Nigeria. Jake did a very fine job explaining what we were doing. Sometimes the Nigerians would get rather upset with us . . . because they have some huge difficulties with Americans that are rather brash. He advised us to approach things in a Nigerian way. He would caution us to be patient and understanding."

As far as Van den Bark is concerned, the Nigerian-AGIP deal might not have succeeded without Simmons's input. "Jake was able to bridge the gap between this American group from Bartlesville and the Nigerians," he says. "We needed his experience and his capabilities with the black race that we didn't have. My opinion is that it would have been questionable that we would have been able to acquire that signature [on the AGIP deal] from the government without Jake. Considering the capabilities that we at Phillips had at that time I doubt we could have done it."[31]

On December 1, 1965, Simmons's work paid off. Alhaji Musa Daggash, who succeeded Sule as oil minister, authorized Phillips Petroleum's license. The company secured an official "working partner" status for 50 percent of Nigeria-AGIP's interests, covering 1.3 million acres of land. Donald Simmons, who was in Nigeria at the time, says that on the night the authorization was signed Daggash invited his father to his home for a private celebration. According to Donald, Daggash said he signed because "Dad wanted him to do it, otherwise screw Phillips and the rest of them."[32]

What Jake Simmons got out of the Phillips-AGIP deal is as indiscernible as his exact role in the negotiations. The documents I read at Phillips Petroleum's headquarters were probably sanitized long ago, as they have been examined numerous times by various government agencies and attorneys involved in lawsuits stemming from failed takeover bids.[33]

The old records at Phillips do contain a few financial agreements with Jake Simmons. On September 24, 1964, Owen Thomas sent him a "consulting agreement" in which Simmons was to be paid $5,000, plus expenses, if the Nigerian government approved Phillips's interest in the Nigeria-AGIP oil prospecting license (he was also commissioned separately to obtain a new oil prospecting license offshore, which is discussed later). On January 10, 1966, Simmons was sent a check for $5,000 for the "acquisition of interest" of the AGIP oilfields. Except for a reference to a $5,000 expense advance, that is all the records show Simmons having made from the AGIP deal.[34]

Either Phillips got a terrific deal on Simmons's work or there is more to it than meets the eye of an inspector of the company's official records. In the risky world of international lease assignments, free-lance concession negotiators are often rewarded for their success with a royalty-like percentage of the profits, a so-called override. The override may be anywhere from a fraction of 1 percent to 5 percent. On a deal which over the past few decades has brought Phillips well over $1 billion in profits (even after paying Nigerian taxes and recouping its investment of more than $200 million), this would have amounted to quite a handsome sum. In Nigeria, such a royalty, off the top of gross production, would have been technically illegal. Any overrides with Phillips, according to an industry expert, would therefore not have been mentioned in public documents. Such restrictions would not apply to the United States, however, so Phillips would be free to enter into a private agreement to pay an override percentage in America.[35]

Edwin Van den Bark says that Simmons was paid his expenses plus a set fee per day to represent Phillips on the AGIP deal in Nigeria, but no override. "We hired him on a per-day basis," he says. "If we were successful we [also] paid him a lump sum. I've forgotten how much. In comparison to today it was rather small, but back then it seemed like a reasonable figure."

When it comes to specific figures Donald Simmons is protective of his family's privacy, especially when it relates to Africa, where he is still involved in the international oil trade. All he will say regarding the AGIP deal is that his father earned a "substantial sum of money" from it.

"Substantial" is a favorite ambiguity of businessmen because it means different things to different people. What "substantial" means in this case can perhaps best be gleaned from something which Jake Simmons told Roy Gootenberg a few years after Gootenberg had sent him on the East African trade mission. "I remember Jake telling me that he owed me something for the fact that I had let him in on the opening to a two- or three- or ten-million-dollar deal—that much which he made out of the Phillips Petroleum deal. He talked in vague terms about it, whether it was a finder's fee or what, but it was no secret." Maurice Keville, the former staff assistant to John F. Kennedy, discussed Nigeria and other deals with Simmons during the early seventies. Although his knowledge is based on personal conjecture, he believes Simmons made far more than a few million from his deal with Phillips in Nigeria. "Jake made more money than anybody who ever went on a trade mission," says Keville.

A portion of this money, however much it was, came from Simmons's other deal with Phillips in Nigeria, which he worked on at the same time as the AGIP negotiations. Phillips was eager to develop its own oilfields in Nigeria, without having to cut its Italian partner in for half. On top of this, the international executives knew that the Nigerian government retained an option to buy into the AGIP production at cost, and that once the wells began flowing its officials would use it. During the late sixties the Nigerians exercised their option. By 1980, Phillips and AGIP's share had dwindled to twenty percent each, while Nigeria's fast-growing Nigerian National Petroleum Corporation enjoyed a sixty percent interest. If Phillips could retain a 100 percent interest in its own oilfield, even one with less overall production could prove more profitable than the AGIP fields. But since virtually all the country's concessions had been allocated before the Bartlesville executives arrived in Nigeria, the odds of them getting their own tract to explore were slim.[36]

Simmons made the difference. Minister Sule gave him a shot at an 897,000-acre concession along the Niger delta. Simmons scooped it up, then "farmed it out" to Phillips. The decision by the ministry to award an individual the concession was unusual, and allowed Simmons the sort of leverage with Phillips that most international oil traders only dream of.[37]

It took more than a year to iron out the details of the Gilli-Gilli field with the Nigerian government. Simmons's compensation was easier to negotiate. Nobody disputes that on this deal he retained an override, probably based upon a direct percentage of gross production from the oilfield. Len Rickards estimates that this percentage was about 2 percent; Edwin Van den Bark recalls it was "probably one-sixteenth" (6.6 percent). Simmons was also paid his expenses and, upon the assignment of the lease, probably a lump sum of $100,000.[38]

The Gilli-Gilli field didn't pan out as well as Phillips's other Nigerian wells. Located about one hundred miles to the northwest of the AGIP fields, the area was the northernmost in Nigeria—too far from the central Niger delta action to share its rich petroleum deposits. In addition, Nigeria's devastating civil war, which began in 1967 and continued through early 1970, disrupted business activities throughout the country. Phillips drilled its first exploratory well at Gilli-Gilli in 1966. It eventually invested tens of millions of dollars there, although it was more than a decade before commercial production began.

By the early eighties the price of oil had topped thirty dollars a barrel, and Gilli-Gilli's annual production of a few thousand barrels daily was worth roughly $12 million a year. Compared to Phillips's one-fifth in its AGIP wells, which produced better than 200,000 barrels a day through 1980 and then trickled down to over 100,000 barrels a day, the Gilli-Gilli field was a drop in the bucket. But it did have an important public relations advantage. With Gilli-Gilli, the company's references to its operations in Nigeria sounded a lot more formidable. "In Nigeria," analysts and stockholders were told, "Phillips has a twenty percent working interest in four licenses and 100 percent in a fifth."[39]

Jake Simmons did not wait around for override checks to make money from the Gilli-Gilli fields. Initial tests showed potential reserves of between ten and twenty-eight million barrels of oil, and there was always the possibility of additional finds in the area. A quintessential trader, Simmons treated his override as a tangible asset and marketed it as such. Between periodic jaunts to Nigeria he contacted associates throughout North America offering to sell them pieces of his piece. Just as he liked to spread the risk on his own domestic wells, he was eager to spread the potential reward from his African deals. The value of the override, before production began, was impossible to measure. It could have increased if exploration proved a better find than expected, or (as in this case) it could have decreased in value if drilling results turned out disappointing. Simmons profited by bringing others in to share in this gamble, turning a speculative interest to money in the bank at the first opportunity.

Simmons milked his override for all it was worth. He was like everyman's oil broker. On a tiny level, he brought about an unprecedented democratization of the business. No longer did an investor have to be a multimillionaire to get a piece of an international oil play. He or she could just buy in to Jake Simmons and join him in reaping the profits from Phillips's expensive exploration program. He sold portions of his override for sums ranging from a few thousand dollars to nearly $100,000.

For a multibillion-dollar company like Phillips, Simmons's marketing scheme was an accounting department's nightmare. By 1971 Simmons's Gilli-Gilli override totaled just 35.7999 percent of 1 percent, estimated to represent about thirty thousand to seventy-five thousand barrels of recoverable reserves. According to Len Rickards an override, in the hands of someone like Simmons, "becomes a merchandisable instrument. He sold it in tiny little fractions. We used to have calls from some of the owners wondering if we were ever going to put the thing into production. And we got so many calls . . . we proceeded to buy up essentially all [but] a very tiny fraction. If it turns out to be a very small amount of money for which there are many diverse owners, the cost of writing the check may be more expensive than the amount of money you send. That's how small it got."[40]

Simmons's experience in Nigeria brought him more than immediate financial gain; it served as an entrée to other African nations. It also indirectly brought his son Donald to enter the arena of African business.

Donald Simmons was the son who most resembled his father, in terms of his oratorical gifts and a persuasive ability to sell people on himself and his projects. He went to work for Phillips's New York office a short while after Howard Edmondson left the Senate in 1964. Later that year, as plans were made to staff a permanent office in Nigeria, Van den Bark tapped twenty-nine-year-old Donald to become manager of sales and development of nonoil products (like gas and petrochemicals) for nineteen independent nations in western and central Africa.[41]

While his father was becoming the first black American to negotiate an international oil concession, Donald became the first black man to run an office of a major oil company abroad.

Donald Simmons got on better with his African associates than his corporate colleagues. He was supposed to report to the head of his division in New York, but clashed with Phillips's exploration chief in Nigeria. It was a classic authority battle for Donald who, like his father, never conformed well to a company mold. "He pretended to be the manager of the whole company there, but he wasn't," recalls Simmons. "He acted like he was the boss of everybody, except he wasn't *my* boss." Ironically the turf battle between the exploration and sales and development divisions in the international department ended up bringing Donald and Jake Simmons into Ghana, the nation that was to become their most important area of operation.

In 1957 Ghana became the first European colony in Africa to achieve its independence. Under the powerful leadership of Kwame Nkrumah, Ghana was the heart of Pan-Africanism, a black nationalist movement committed to wiping out every manifestation of colonialism and bringing about the eventual unification of all Africa into a single nation.[42]

By 1965 Nkrumah was looked upon as a world-class leader—and a difficult man to do business with. Multinational oil companies, which were spending billions exploring for oil in Nigeria, just a hundred miles to the east, were hesitant to venture into Ghana. Nkrumah knew this. He also knew that the financial squeeze brought about by his ambitious Pan-African projects would never allow Ghana—a fraction the size of Nigeria—to finance its oil exploration by itself. When reports of the billions of dollars annually which Nigeria expected to reap from its petroleum harvest reached Ghana, they softened Nkrumah's antipathy for the evil oil cartel. He ordered Sylvan Amegashie, his country's first CPA and chairman of the Capital Investments Board, to do what was necessary to interest the oil barons in his country.[43]

In early 1965 Donald Simmons sent his Nigerian assistant, an MIT-trained engineer named Muili Salami, to Accra, the capital of Ghana, to investigate the overtures Amegashie's Capital Investments Board was making to the international business community. Salami met with Amegashie, who encouraged Phillips to make a play for Ghana. During the course of their conversation Salami explained that although he worked for a major American oil company, his boss was a black man.

Amegashie, a gregarious man, liked a good joke, and he thought he was hearing one. "You're full of shit," he said. Salami insisted he was serious. Amegashie suggested that he bring his boss with him on his next trip to Ghana.

Simmons returned with Salami a few weeks later. He and Amegashie hit it off right away. They had a lot in common. Both were successful businessmen with strong appetites for food, drink, bawdy jokes, and good stories. Before the trip was over Simmons was meeting with Amegashie in his home. A fraternal relationship between the two was formed which continues, along with mutual business interests, to this day.

Amegashie was impressed by Phillips's reputation and its financial muscle. British Petroleum had done some light preliminary work in Ghana years earlier, but had been unwilling to become committed to any real exploration there. The only drilling which had taken place had been a disastrous attempt by the Rumanian government in the early sixties. The Rumanians brought in second-rate equipment and stuck the Ghanaian government with the tab—more than $4 million in precious foreign currency.[44]

Nkrumah may have raged against Africa's dependence on international capitalism, but only international capitalists were willing to invest the kind of money and expertise Ghana needed to develop its potential petroleum reserves. Amegashie was authorized to hire a British consultant to conduct a few seismic tests and assemble the basic geological data needed to woo a multinational. If interested, the foreign oil company would risk its money (simulating the local economy in the process) and Ghana could share in half the profits. There was a difference between exploitation and a mutually beneficial partnership. Ghana needed a partner.

Donald Simmons rushed back to his Lagos office and sent a telex to one of the heads of the exploration division back in the United States. Ghana was wide open, he wrote. The government was ready to negotiate exploration rights for virtually the entire country. What did they want him to do next?

A terse rebuke from an executive in the international exploration division informed him that he was overstepping his reach. Finding oil was not what he had been sent to Africa to do. "You're not in the exploration division," he was told. "You're in sales and development. We can handle our own matters."

Donald Simmons had done his job for Phillips. Next he did his job for Simmons. "I said, 'Fine, thank you,' " he recalls. "Then I called my old man and told him that we could get an oil concession in Ghana. It was later that we got millions of dollars in bonuses from Phillips just to come in and take a portion of it. So things have a way of getting poetic justice."[45]

Jake Simmons had done a good job training his children not only to spot opportunities when they arose, but also to create opportunities whenever they might be conceived. Donald's father had taught him that the capitalist equation meant making such connections; that the worker's labor is designed to create a profit for his boss, but that the laborer, at times, has a chance to use the affiliation of his employment for his own education, or advantage. This did not mean being disloyal to one's employers. It simply meant using them, just as they were using you. Jake had done this with the U.S. Department of Commerce trade mission, the Nigerian government, and Phillips Petroleum. Like a bee cross-pollinating flowers to allow them to flourish, Jake came away with a good portion of honey in Africa by connecting various groups for their mutual benefit, always using his association with one to score points with the other. To this day Donald Simmons, before he enters into a business deal, insists on satisfying himself that he understands how the other party intends to profit from the transaction. "If I hear anybody telling me they've got a deal that's real good for me but bad for them, I say, 'Look out, you're about to get your pants taken off.' My father used to say whenever you hear anybody talking about Communism they're not talking about but one thing: *you* work and *they* will reap the benefits from it."

Jake Simmons knew that his race, in Africa, afforded him certain advantages (what Donald Simmons refers to as "a greater receptivity"), but he also knew that the color of his skin would only take him so far. As he instructed his sons, "Just being black is never enough. You have to have competence behind that blackness."[46]

To this day Owen Thomas regards Simmons as an exceptional role model. "I talk to my family a lot about Jake," says Thomas. "In spite of coming into a world where the cards seemed to be stacked against him he was strong enough to overcome those obstacles and win the respect and admiration of the people in the system who worked with him. The fact that he could experience what all black people experienced in those days and still not be bitter was a remarkable achievement."

Simmons's African associates were no less impressed by his character. As a corporate diplomat, or what the National Petroleum Council, in a memorial resolution, referred to as "an extremely effective petroleum industry statesman," Simmons won personal admiration, respect for Phillips, and respect for black Americans in general.[47] "He made a lot of friends," explains Henry Aliva Johnson, an internationally successful Lagos-based businessman. "He was a man we Nigerians turned to at different times in our lives. I asked advice of him often during my matrimonial problems. He told me how I should approach things and every word he told me has come true." On a professional level, Johnson adds, "He made a great impact promoting the image of Phillips in Nigeria. His public relations was just wonderful." Others saw the Oklahoman as promoting far more than himself and the company he worked with. Former Nigerian ambassador Maitama-Sule said that Simmons "conducted himself as a very good ambassador of the United States and of the black people of the United States."

Jake Simmons's accomplishments are even more remarkable when placed in the context of the tempestuous times in which he worked in Nigeria. The country is a union of some three hundred ethnic groups and languages. About 60 percent of the population belongs to one of three regionally dominant groups: the Hausa-Fulani in the north, the Yoruba in the west, and the Ibo in the east. Attempts to balance the representation of all three groups in the federal government failed dismally during the early years of Nigerian independence, inspiring allegations of widespread favoritism and voting irregularities.

In early 1966 a group of Ibo army officers assassinated the government's leaders, ostensibly to restore order. A countercoup soon followed, along with a bloody purge against the Ibos. More than one million terrified Ibos fled from the north and west for the safety of their eastern homeland. On May 26, 1967, the Ibos seceded from Nigeria and declared the independent Republic of Biafra. For the next thirty months the nation was plunged into a tragic civil war.

The turmoil derailed negotiations which Jake Simmons had been conducting for another Phillips concession. It also killed Donald's prospective projects for the company. Yet Africa was much larger than one country. Millions of acres of unallocated land awaited exploration in Ghana, a country which looked better and better as the conflict in Nigeria worsened. By mid-1967 both Simmonses had left war-scarred Nigeria, anxious to take what they had learned with Phillips and use it to stake out new territory for themselves.[48]

In 1967 Simmons began traveling to Ghana even more frequently than he had traveled to Nigeria in earlier years. Ghana had gone through a dramatic transition within a short time of Jake Simmons's first visit there. Early on the morning of February 24, 1966, while strongman Kwame Nkrumah was in Hanoi on a mission to bring peace to another part of the world (he linked Africa's fortunes to the "worldwide struggle against imperialism"), a large group of Ghanaian army officers and police staged a military coup.[49] To its credit the new National Liberation Council, led by Lieut. General Ankrah, lived up to its promise of allowing free and fair elections by 1969. In 1970 it turned over all authority to a civilian government, becoming one of the rare military governments in African history to voluntarily relinquish power. (Democracy was short-lived; within a few years a different military government seized power.)

General Ankrah's government was even more eager than the one which preceded it to see Ghana capitalize on its oil resources. Sylvan Amegashie was a central figure in making this happen—first as part of an eleven-member national economic committee, then as commissioner for industries, and finally as commissioner for lands and mineral resources.[50]

Amegashie viewed Simmons as the man who could link Ghana with the oil multinationals, and also help him figure out how to establish a domestic exploration program. He forwarded geological information to the Oklahoman, and met with him, both in the States and Ghana. Simmons examined the prospects and worked for an opportunity to angle himself into a deal.

First he needed to make a connection with a source of big money. This, Simmons knew, would elicit greater cooperation and allow him to leverage his influence with the Ghanaian authorities. He had heard that Los Angeles–based Signal Oil & Gas Company had started what its officers called "an aggressive mode" to replace reserves with new oil finds. Backed with an annual commitment of $60 million for exploration and development, Signal was eager to initiate major international deals. The company had been unable to win an oil exploration concession in Nigeria and was considering settling for a gas processing deal there. When word of this reached Simmons, he approached Signal with a far more attractive offer.[51]

Simmons knew no one at Signal, but that did not stop him from marching into the company's headquarters in late 1967 with a map of an enormous concession in Ghana which he claimed to have a hold on. He was routed to the company's chief geologist, Loring B. Sneden, who was somewhat skeptical at first. Having anticipated this, Simmons had brought an important reference. He told Sneden he could check his qualifications with a Texaco land executive from Tulsa who he knew had worked with Sneden in the past. Sneden called his associate, who told him of satisfactory deals his company had made with Simmons in the past. He tells things the way they are, the executive explained, although "as a land man he's an optimist."

The recommendation got Simmons into Signal's executive suite where he wasted no time in becoming buddies with the top brass. "He just never met a stranger and he wasn't about to be ignored," recalls W. H. Thompson, then Signal's president. "He recognized that the people who were going to put big dollars into big oil were not going to be looking for forty-acre prospects in Creek County, Oklahoma. They were going to be looking for billion-barrel oilfields on the continental shelves of the world. The trick was to get himself situated so he could get between the reserves and the money. And that's exactly what he did."[52]

By this time Jake Simmons had familiarized himself with the lingo and logistics of international oil concessions, and sounded as though he had been handling them all his life. He was able to reel off the minute details of a contract from memory. Signal president Thompson, a geologist by training, recalls that Simmons's command of geological terminology and concepts was so advanced that he "could talk and you would guess he had a degree in geology or geophysics."

His persuasive abilities were no less impressive. Soon after meeting

Simmons for the first time, Loring Sneden found himself on a plane bound for Ghana to see if the Oklahoman had the kind of influence he claimed to. Simmons also had to allay the conservative geologists' fears that the radical days of Kwame Nkrumah's regime were a thing of the past, and that the Ghanaians were "mending their ways."[53]

After meeting with Sylvan Amegashie and Ghana's chief geologist, Sneden reported back to his superiors that Simmons's offer was for real. Furthermore, the preliminary geological data provided by the Ghanaian government revealed a number of potentially gigantic oilfields. It was clear that here was a chance for Signal to land a major offshore concession along West Africa's continental shelf.

It was also clear that commercial quantities of oil had never been discovered in Ghana, so the $15 million or so which it would cost to carry off a competent exploration program would be entirely at risk. Signal was ambitious, but it was no Exxon Corporation. Its executives contacted associates at Occidental Petroleum, a California-based company it was working with on other foreign projects, and invited them in on the deal. At some point in the negotiations, the partners were joined by Amoco, a subsidiary of Standard Oil of Indiana, which was also eager to invest in an international "play." Signal was to be the operator in charge of the exploration program.[54]

Jake Simmons acted as spokesman for the partnership's five-man negotiating team, which consisted of an attorney and a ranking international executive from both Signal and Occidental. Through 1968 the group traveled to Ghana nearly every month to negotiate with government officials. After each trip the oil company representatives had to get approval from their higher-ups authorizing them to take the next step. Working up a contract with Amegashie and Ghana's attorney general Robert Benjamin was not as easy as adapting the terms of a standard concession agreement. The Ghanaian Ministry of Lands and Mineral Resources had never done this before. The Signal contract would be its first ever, and in developing it the Ghanaian authorities had to develop an all-new petroleum code, one which they would use for other companies vying for concessions in the unexplored country.[55]

To Sylvan Amegashie, Jake Simmons was more than a contract negotiator sitting across the table. In Amegashie's important two-year tenure as commissioner of land and mineral resources, his country was opened to exploration by some of the world's largest oil companies. Yet when he started out he had no idea even how to draw up a petroleum concession, much less cut a decent deal with highly experienced international executives. Amegashie needed advice, and he needed it from a source he could trust. Jake Simmons became that source.

"All my advisers at the time," remembers Amegashie, "were professional arm's-length people whom I hadn't known before. Now here was a very respectable, experienced black oil friend, the father of Don Simmons. We asked him to advise us because he was one of us. I found that I could take him into confidence to educate me in all the things I needed to know before my talks with all the other people. To me, he was one hell of a find. I became virtually his son. The man was my mentor."

Simmons gave the young commissioner (who was not yet forty) books on mining and oil terminology. Although unpaid, he became the ministry's most important consultant. After intensive negotiations, Amegashie would send a car to pick up Simmons and bring him to his home to discuss the merits of the deal that was being offered to Ghana. Simmons's associates from the Signal consortium were eager to work out a fair arrangement, to ensure that an overly advantageous deal would not "cause trouble down the line." They had no trouble seeing the benefit of allowing Simmons to play both sides of the fence. The Oklahoman became indispensable, advising Amegashie on legal and contractual terms, and pointing out how Ghana could best ensure that the oil companies carried out their obligations.[56]

Simmons was very much at home in Ghana. Unlike his American business associates, who spent most of their time around the hotel and were eager to head home after a few days of negotiations, Simmons lingered and established a network of social contacts. His social accessibility made him a favorite of Accra society, and he rarely went without lunch or dinner invitations.[57]

One of his closest friends in Accra was an African-American expatriate ironically named Robert E. Lee. Lee, like Simmons, was a highly articulate man with powerful political passions. Malcolm X referred to Lee and his wife—both of whom are dentists—in his 1965 *Autobiography* as "militant former Brooklynites who had given up their United States citizenship." They are among the most prominent members of a sizable community of African-Americans who live in Ghana. Lee studied with Kwame Nkrumah at Lincoln University. His leftist view of African independence and his virulent anticolonialism (which includes a strong anti-American sentiment) might have made him an unlikely associate of the intensely patriotic Jake Simmons in the United States, but in Ghana the two men saw eye to eye.

I met Dr. Lee on a breezy patio near downtown Accra in the summer of 1986. Although in his seventies, he remains remarkably youthful-looking, a wiry man who was interviewed wearing a maroon jumpsuit. Smoking a pipe, he managed to appear both distinguished and relaxed. Lee always regarded Simmons as a symbol, he said, "of what it could be like to be an ordinary, intelligent, industrious, civic-oriented, politically aware American no matter what color he is. I saw him as a kind of new black American. My experiences growing up as a black American told me that blacks always had two faces: himself—whom you rarely saw, and the person inside a kind of racist organization. You never could tell what kind of personality you were actually talking to. But Jake Simmons didn't have any complexes. He wasn't afraid of anybody and he was truthful, he was himself. He didn't have any chips on his shoulder. He was comfortable not only being black, but being an ordinary American."

Lee noted that Simmons evoked a similar confidence from the Ghanaians. "Most of these new African states were secretly hoping that they would have oil struck on their territory, but they were holding back because they felt the big oil magnates of the world would just use up their oil and they wouldn't get very much out of it," he explained. "Simmons gave them a bit of relaxation, knowing he was a black fellow who knew something about oil. They listened to him and got some guidance as to how to go about making those agreements."

Simmons always notified Dr. Lee and a clique of a half dozen other Ghanaian-Americans whenever he arrived in Ghana. They would throw a small reception for him, eager to hear what was happening in the States. Simmons was just as popular with his Ghanaian friends, and when they visited America, he made sure they got out to Oklahoma.

Such camaraderie also extended to Simmons's Ghanaian business contacts. At eight o'clock one morning in 1967, Sylvan Amegashie, then commissioner of land and mines, was fast asleep after staying up until 5:00 A.M. finishing some important paperwork. He had to deliver his work to the Ghanaian cabinet at ten-thirty, and had instructed his wife not to wake him up, no matter who came calling. At eight Jake Simmons arrived at the house, needing to discuss some pressing business. He marched right past Mrs. Amegashie into Sylvan's bedroom, and shook him awake. "Get up, sonny," he said, and repeated one of his favorite expressions, "No black man has any business being in bed when the white man is running the world—get up!"

Amegashie seems as amazed by the incident today as he was twenty years ago. "Can you believe that?" he asks. "Somebody walks into the bedroom of a minister of state and tells him that! Jake was the only one who would ever do something like that."

Amegashie knew full well that Simmons's main goal was to make money. But this did not mean he could not help Ghana in the process. When talking to associates like Signal geologist Loring Sneden, Simmons explained how much an oil find would mean to Ghana's struggling economy, and convinced him that their effort in developing an equitable development program was a noble one. "I think to begin with his interest was purely professional," explains Amegashie, "but as time went on it became fraternal. That he was going to benefit was obvious. But he soon realized that with his help we could also work to develop this part of Africa. He wanted Africa to come good, and he wanted to earn a living. There's no reason the two things should be in conflict."

After a lifetime of deal-making, Simmons had created a unique role for himself. He was in a position to do tremendous good for a hard-pressed African people by doing tremendously well for himself. The advantage of his status was not lost to his competitors. Independent broker Marvin Billet, a veteran of numerous international oil concessions across West Africa, was playing a middleman game in Ghana similar to Simmons's in 1968 for his company, Aracca Petroleum. "Jake was somebody the Ghanaians felt they could trust and he would not do the wrong thing by them," Billet recalls. "Africans, once they feel that way, are very loyal. He acted ex officio, sort of as their adviser. It's a marvelous position to be in."

Although the performance terms of each oil company's agreements were similar, Amegashie's ministry made the ultimate choice of which concession "block" each contender would get. With Simmons's advice the ministry was shrewdly following the path of competitive participation. Instead of relying on one or two corporate giants, Amegashie divided the entire country into concession blocks and attempted to attract enough oil companies to ensure that every region of the country would be explored. Not all blocks were equal, however. It was Simmons who made the selection for the Signal consortium. He was reportedly allowed to pick what he believed to be the cream of the crop. From an offering of more than twenty-two available parcels he chose four blocks in the ocean off the coast of the small port city of Salt Pond. The area amounted to about one million acres, much of it along the continental shelf. Early tests looked promising. Signal's former president W. H. Thompson said, "We tended to think of it as a potential Nigeria situation. We were looking for potential billion-barrel oilfields there."[58]

But the Signal deal could not be completed until the final terms were agreed on, and one major difference threatened the deal. Encouraged by the number of interested oil companies, Amegashie and Simmons had worked out a stringent performance schedule which allowed the Ghanaian government to make money every step of the way.[59] But early in the negotiations Amegashie also asked for an up-front windfall of $10 million to pump into the Ghanaian economy. Simmons, knowing how important such an infusion of foreign capital would be to a government that operated on a budget of less than $350 million, suggested that the Signal group satisfy the ministry by paying the government a few million.

Signal's management at first refused to pay anything, insisting that they wanted to apply such money toward exploration. The whole project nearly fell apart. It was the only point on which Signal disagreed with the Oklahoman, and the one area in which its executives wondered whether Simmons's dual role might have stretched his loyalty too far. Eventually the middleman smoothed over the difference, convincing the Ghanaian authorities to accept what Loring Sneden called a far smaller "token" bonus, and convincing Sneden and his superiors that he would make up the difference between what they wanted to pay as a bonus and the minimum which the Ghanaian government would accept out of his own pocket. (How much this was and whether he actually ended up paying a share of it is unknown.)[60]

Even if Simmons had to make a nominal contribution toward a

final agreement, it would have been worth his while. As is common in such arrangements, Signal's contract with the Ghanaian government was a multistep deal. The further along the road to profitability the Signal group went, the more Simmons would make. The first step to the deal was a "seismic option"—authorization to bring in heavy equipment and drill exploratory wells. Simmons received about $100,000 at this point, and another $100,000 if the consortium chose to follow up with a "winning," or operational license (which it did), permitting it to extract and export oil. Simmons also retained a 4-percent override on all the group's income (after costs), and various bonuses at different stages of production. On top of all this, Simmons worked in a provision stipulating that any part of the concession which the consortium relinquished, along with the data resulting from its multimillion-dollar exploration program, would be reassigned to him personally.[61]

Amegashie was determined to have his country's drilling concessions distributed by the end of 1968. He planned a large contract-signing ceremony on December 19 at the State House in Accra. It was to be a significant milestone in Ghanaian history. Sylvan recalls that he and his countrymen were "hoping that we were going to be the next big oil people." Plans were made for a major affair, observing all the formalities of international protocol. The entire Ghanaian cabinet, the diplomatic community, and the international media were to be present. Everything seemed to be going smoothly when, on the evening of December 18, Charles Adams, the U.S. Embassy's economic attaché, visited Amegashie's home. Adams, a friendly middle-aged white American who knew Sylvan well, had bad news. "I've just come from a meeting of the American oil companies," Amegashie remembers Adams telling him. "The consensus was that your terms [the performance obligations] are too rigid. I'm sorry to say you're going to be very disappointed tomorrow because not a single one of them are going to sign. I suggest you postpone the meeting and renegotiate."

Sylvan had been bargaining over terms with the various oil companies for months. The contracts had been worked through and he expected that getting the executives to sign would be little more than a formality. He tried to sound sure of himself when he told Adams that he was confident the signing would come off as planned. Adams was ready to bet him a thousand dollars it would not. Amegashie declined, but was willing to bet a bottle of champagne.

Despite his dauntless front, Sylvan was worried. Among the companies which had converged on Accra were massive multinationals like Gulf, Exxon, Texaco, and Mobil. "It was rather startling news," Sylvan recalls. "It struck me that the major companies were ganging up to act as a block and ditch the contract. We were on the threshold—this was to be our breakthrough. If they blocked that signing it would have been a major disaster."

Amegashie was holding just one ace: his faith in Jake Simmons. "Jake has to sign," Amegashie told himself as Adams left his home, "because he was virtually the author of the agreement." Sylvan telephoned the Oklahoman. "What's this I hear about the Americans forming a conspiracy to break up this thing?" he asked.

"Rest assured," Simmons replied. "Our consortium will sign."

The next day, scores of flags from around the world circled the Ghanaian State House as dignitaries streamed out of their Mercedes-Benzes to witness the beginning of Ghana's oil era. Amegashie joined Ghana's other economic ministers at a dais in the front of the conference room. All wore Western-style suits. Television cameras purred as Amegashie gave his opening remarks, then invited the corporate representatives present to join him at the dais, present their letters of accreditation, and sign the prepared contract.

Jake Simmons stood up immediately and paced hurriedly to the front of the room. Instead of signing right away, he took the opportunity to make a speech. "Ghana is a country struggling to emerge from underdevelopment," he began, "and the rest of the developed world should make it its business to respond to this." He talked more about the nation's oil potential, concluding that as an American of African descent he was proud to sign the exploration agreement on behalf of Signal, Occidental, and Amoco. His corporate partners also signed. They were followed by a representative for a subsidiary of Standard Oil of California. Jake Simmons had broken the ice. A celebration buffet followed and that evening Charles Adams again visited Amegashie's home, this time with a bottle of Dom Pérignon.

It did not take long for the holdouts to buckle under. A second ceremony took place twelve days later, on December 31, 1968. Amegashie had given the uncommitted oil companies a few weeks to realize that he would not alter his terms. Representatives for Mobil, Texaco, and the Israel National Oil Company came forward and signed their agreements.[62]

With the commencement of exploration by more than a half dozen multinational groups, Simmons's role as broker and unofficial government consultant diminished, and he entered a new phase of business. As he had done in Nigeria, he divided up his 4-percent override stake and began selling bits and pieces of it off. They did not come cheap. Signal geologist Loring Sneden anticipated a find in the hundred millions of barrels, although he notes that his company's top execs were often more optimistic in their estimates and talked about a billion-plus find. It is likely Simmons joined the optimists in making his sales pitches.

How much Jake Simmons was able to get for such potential would have depended on whom he was selling it to. (After all, it was an oilman, Shell founder Marcus Samuel, who in 1911 coined the expression "The price of an article is exactly what it will fetch.") While frugal enough to cash in a stack of his chips before the crapshoot was over, Simmons was also enough of an entrepreneur to play the upside potential: he is thought to have retained nearly half of the 4 percent for himself. His ability to market percentages of his percentages was bolstered in mid-1970 by the Signal consortium's initial well, which flowed at nearly four thousand barrels a day. Simmons's concession selection seemed to be a good one: Signal was the first operator to find oil in Ghana. "The reason I was optimistic," explains Loring Sneden, "was that the structure looked big, and even if it wasn't a 100,000-barrel well, we could get one hundred 5,000-barrel wells and that would do it."

Signal drilled four more wells, and none came in as strong as the first. The test wells provided the company's geologist with disheartening news. The large closure which they hoped contained a billion barrels of oil was riddled with faults, causing much of the oil which may once have been there to leak out. There was no way to determine where oil could best be found, which meant every new well would be a blind shot, or "wildcat." The offshore drilling cost too much to warrant the odds. After an investment of about $15 million, Signal, like most other oil companies in Ghana, called it quits. They abandoned the concession, which, according to their agreement, meant that all the rights to explore the area, and all the data they had painstakingly assembled, suddenly belonged to Jake Simmons.[63]

Simmons continued to do business in Ghana, peddling a number of concessions he controlled to groups led by both Texaco and Phillips Petroleum. He received large payments, plus overrides from both of these deals.[64]

Just five years before he died, Simmons, with help from his son Donald (whose African oil business continues to this day), assembled one of the largest deals of his life. By 1976 he was about the only broker who had not given up on Ghana. At the time there had been no exploration for three years. Some forty dry holes had been drilled, costing a variety of companies more than $100 million. Yet none had discovered truly commercial quantities of oil. Ghana's dismal geological track record, Simmons realized, meant that the Ghanaian government would be willing to offer far more generous terms on its concessions than it had in 1968. He maintained the right to explore the blocks he had farmed out earlier, and possessed valuable seismographic information. He just needed to get a big-bucks backer. He turned to one of America's newest and most unusual oil exploration companies, Agri-Petco International.

Tulsa-based Agri-Petco was controlled by a group of middle-American farm cooperatives. The cooperatives owned three refineries in the Midwest, and sold refined products for agricultural use directly to their farmer members. After the Arab oil embargo of 1973, the major oil companies which had supplied the refineries with crude oil shut them off. The coop members decided they would pool their resources, form an exploration company, and find oil themselves. They dabbled in small projects in Egypt and Greece, then were persuaded by Jake Simmons to dive full force into Ghana.

Once again, Simmons traveled to Ghana to negotiate the deal. A military government had seized power from Ghana's democratic leadership in 1972, and experienced appointees like Amegashie were no longer in office. The first oil concessions in years were handed over to the newly formed Ministry of Oil and Power. Michael Akotu-Sasu, at the time a bureaucrat with no knowledge of the oil business, was given the responsibility of dealing with Simmons. Once again, the Oklahoman assumed the role of trusted adviser. "I just happened to jump into it; he could have exploited my ignorance, but he didn't," says Akotu-Sasu, who despite three additional changes in leadership is still a Ghanaian government official. "He just took me as some sort of son, and tried to advise me how to do my new job, which was very complicated. Whenever he felt that Ghana would lose he would point it out to us. He wasn't out to cheat us—he was more concerned that Ghana should get oil to help us with our economic problems. He felt that just like Israel, which Jews from every country can look up to and go to, there should also be a black African state that has also come out of the Third World, which an American or British or Jamaican black can go and feel comfortable in, and point to as a nation that has succeeded."

Simmons, while making sure that Ghana got its due, also struck an excellent deal for Agri-Petco. In late 1976, he secured for them the same Salt Pond acreage which Signal had nearly succeeded with. It was to be the most extensive exploration program ever undertaken in Ghana.[65] If that billion-barrel offshore formation existed, Simmons was determined to find it.

9

A SEAT AT THE TABLE

Money doesn't talk in this country, it screams. Politically we had to have some money or friends with money. It was very important that Jake had money. He did his part not as a quarterback in the sit-in movement, but a blocker. He was up there blocking and opening doors that we didn't even know about.[1]

—Clara Luper, Oklahoma City schoolteacher and organizer of the nation's first lunch counter sit-ins

Although Oklahoma never received as much attention as states like Alabama and Mississippi, it was at the forefront of the struggle for racial justice. Numerous historical accounts incorrectly refer to a February 1, 1960, sit-in at a Woolworth lunch counter in Greensboro, North Carolina, as the start of the nation's integrationist sit-in movement. In fact, it was in Oklahoma City that the first effective program of lunch counter sit-ins began. The city's NAACP also achieved the honor of carrying out the longest-running nonviolent sit-in campaign ever. Beginning in August 1958, groups of integrated schoolchildren occupied every kind of business establishment and public accommodation imaginable, from lunch counters to amusement parks and church halls. Led by Clara Luper, an articulate schoolteacher, members of the Youth Council's well-organized action squads spent more than six years tearing down Oklahoma City's deeply rooted walls of segregation. As they waved American flags and sang hymns, they were threatened, spat upon, assaulted and arrested by the dozen. At one point members of the state legislature poured human feces and urine from a balcony onto a group of protesters at the state capitol. The activists were relentless, integrating one facility after another. Most mainstream Oklahoma NAACP leaders criticized their tactics. Jake Simmons gave them his full support.[2]

"Simmons was totally different from the black leadership in Oklahoma at the time," reflects Edward Melvin Porter, who headed Oklahoma City's NAACP from 1961 to 1964. "Most of the confrontations with the white power structure at the time were done by young people. The older leadership was a stumbling block. They felt that white people should be dealt with differently than through direct confrontation. I was called into a meeting with fifteen of the black leaders here who told me we were just wrecking the city. That was the only time I ever cried over my involvement in the movement. Jake Simmons was just the reverse of that. He was articulate, economically secure, and militant. He asked for no quarter and gave none. I think he would have died a violent death before he would have subscribed to a white system reducing him to a submissive form of living. He was fearless. He motivated you to never think of fear."

Porter, an attorney, spent much of the early sixties defending young civil rights protesters. He ran for the state legislature in 1960 and 1962 and lost both times. In 1964, when federal courts demanded that election districts be reapportioned to allow black districts for the first time, it became clear that Oklahoma City blacks would be able to vote an African-American into the state senate. Porter threw his hat in the ring. In response the city's power structure put their support and money behind one of the black community's most proestablishment leaders, F. D. Moon. Moon had the combined weight of Oklahoma City's Chamber of Commerce, the Retail Merchants' Association and the *Daily Oklahoman* behind him; Porter raised loose change at cookie sales. The largest contribution Porter received, for $150, came from Jake Simmons. He won the election and for more than two decades was Oklahoma's ranking black politician.[3]

As chief fund-raiser and later president of Oklahoma's NAACP, Simmons was an important ally of Porter's main clients—Clara Luper and her thousands of "kids." Simmons knew that every victory in Oklahoma City would encourage similar protests in cities like Tulsa, and set examples which other businesses around the state would be forced to follow. The war to integrate Oklahoma City, by Clara Luper's reckoning, cost more than $100,000 (the Youth Council groups also traveled around the state assisting other youth chapters). As one of Oklahoma's most prominent NAACP leaders (and, after 1962, the state president), Simmons solicited donations from anonymous businessmen and gave generously himself. "Money doesn't talk in this country, it screams," explains Luper. "Politically we had to have some money or friends with money. It was very important that Jake had money. He did his part not as a quarterback in the sit-in movement, but a blocker. He was up there blocking and opening doors that we didn't even know about. I think his success was in what we didn't see, the contacts that we didn't know about, the telephone calls that he was able to make. I have stood down there on the street many a day and someone walked up to me that I didn't know and said, 'I'm a friend of Jake Simmons's—I want you to have good luck.' "

Clara Luper's favorite Jake Simmons story concerns the single time she was able to convince the NAACP chief to join her group for a nonviolent protest.[4] It was in late 1964, and the Youth Council was preparing to demonstrate in the lobby of the Skirvin Hotel. Known as the "Waldorf-Astoria of Oklahoma City," the Skirvin was one of the last vestiges of local segregation. As the hierarchy of the state's NAACP were concluding their annual gathering, Luper and her supporters marched into the meeting room and demanded that all the NAACP leaders join them in a nonviolent protest at the Skirvin. Jake and the others agreed to accompany the group, then listened to Luper explain the principles of nonviolence which her groups had upheld for more than six years. "If someone spits on you, smile," she instructed. "If someone curses you, smile." Simmons looked uncomfortable but said nothing. As the group walked silently through downtown Oklahoma City to the hotel, Luper had her doubts.

"The only thing was," Luper remembers, "Jake was no backup fellow, so he was going to lead. We walked into the Skirvin Hotel and as we stood in line, a big burly white man came and pushed Jake Simmons. At that moment he proved to us beyond a shadow of a doubt that he was not a nonviolent man. I have never seen anybody as angry as Jake was, and though he was very devoted to the cause he called that man everything he could think of. Sweat was popping out of his skin like popcorn and dropping to the floor. I intervened and called Jake aside and told him that he had flunked his test in sitology. He handed me twenty-five dollars to buy the kids lunch and said, 'I'm not cut out for this kind of thing. I'll kill that man. I have to leave.'" Luper was relieved to see him go.

Simmons found nonviolence unacceptable—even for his sons. In 1963, after watching Sheriff "Bull" Connor turn a pack of vicious dogs loose on a group of helpless black demonstrators, many of them women, in Birmingham, Simmons angrily warned Donald, "If one of you lets a cracker turn a dog loose on your mother or your wives, don't bother coming home!"

Although Simmons did not have the disposition to man most picket lines himself he was quick to use the threat of peaceful demonstration to push through the NAACP's civil rights agenda. As early as 1960 he was able to convince Muskogee's largest hotel and most of its restaurants to integrate.[5] "Jake Simmons was smart enough to use Oklahoma City to open up Muskogee," recalls Clara Luper. "He moved in a class of people like the Chamber of Commerce, people that really run a city. And he told those people, 'We don't want to upset things in Muskogee. We don't want any sit-ins over here and we don't want any marches. Let's get together and do it.' He just used the threat of Clara Luper and her kids coming to town. And consequently in Muskogee we did not have one sit-in."

Simmons's power was felt most keenly in his own community. He took a leading role in the region's civil rights struggle, especially in the eastern half of the state. He was so influential in Muskogee County Democratic politics (he was secretary-treasurer of the party for two decades) that although the district was three-quarters white, he was chosen in 1960 to be Muskogee's sole delegate to the Democratic National Convention in Los Angeles. Two other blacks accompanied the Oklahoma delegation to the July 11 convention, but as alternates. Simmons, who also attended party conventions during the fifties, was one of the first black delegates Oklahoma ever sent to a national political convention.[6]

The contest for the party's nomination was between John F. Kennedy, the charismatic senator from Massachusetts, and Lyndon Johnson, the powerful senator from Texas. Simmons believed Johnson would be more effective than Kennedy on racial issues. Both candidates had established only modest credentials on key civil rights legislation while in the Senate, yet it was Johnson whom Simmons trusted more to do the right thing for African-Americans. J. J. Simmons III recalls being "a little miffed" when he found that his father backed Johnson. "The man is a man of character," Jake Junior explained to his son. "Do you realize that most of the monies we have made in our lifetime have been made with people that many would have considered reactionary, and that some of our best friends have at some time belonged to adverse groups? You have to judge the man, the total man, rather than his reputation."

To members of the white establishment, Simmons's reputation as a successful oilman enhanced his credibility as president of the Oklahoma NAACP. Even the *Daily Oklahoman,* a bastion of Oklahoma City conservatism, allowed its pages to be used as a soapbox for Simmons's beliefs. In the only profile of Jake Simmons ever written, the paper's Sunday magazine, *Oklahoma's Orbit,* ran a two-page article about the state's NAACP president on August 25, 1963. Entitled "When Men Forget Color," the story attempted to paint Simmons as a moderate, although the views he expressed to reporter Bill Harmon were far from complacent. He leveled harsh criticism at city and state officials who were "dragging their feet." Playing on the phobias of the day, Simmons said:

> Integration is not an accommodation to colored people. Giving all Americans their rights will magnify the United States' position in the world. The most serious threat that democracy has today is segregation and discrimination. My contention is that discrimination is more serious than Communism and socialism. Here is the economic core of it again. Men must have a chance to work to make a living, or they will turn to socialism to answer their needs. The clock is ticking fast on America as the last important stronghold of the democratic system. We cannot allow segregation, injustice and discrimination to weaken it.

Simmons may have found sit-ins difficult to participate in personally, but he had no compunction about advocating their vigorous use. "The nation cannot afford to let time solve the problem," he said. "Negroes must press now for their rights by all legal means and by peaceful demonstrations. . . . We have men in public office who subscribe themselves to discrimination in order to perpetuate themselves in office. Politicians in the South have played on prejudice, have played Negro against white. We should get rid of such men now."[7]

Doing it now was one of Simmons's favorite themes. When it

came to social injustice, he had little patience for slow change. As president of Oklahoma's twenty-four active chapters, the theme of the first statewide conference which Simmons headed, in November 1963, was "Full Citizenship Rights NOW. Moderation LATER." He pushed hard for full school integration, a contentious issue among some black leaders who correctly predicted it would cause the wholesale firing of black teachers when segregated schools were closed. Simmons threatened to get the NAACP to persuade the federal government to withhold funds from school districts that discriminated against black teachers. "Negro teachers have suffered severely," he told the *Orbit* in 1963. "We are going to serve notice on every school board in Oklahoma to adopt a fully integrated policy in hiring teachers. And we are going to set a deadline for 1964." The first school board Simmons "served notice" on was Muskogee's. Simmons also wanted to integrate the bookshelves of the Muskogee schools. He later contacted an official at the NAACP's national headquarters in New York asking for a list of books and periodicals by African-American authors. "We have recently dedicated a new high school in this city," he wrote, "which is fully integrated. I will not be satisfied until every worthy periodical concerning our race is placed on the library shelves."[8]

As Simmons's political influence and the issues he stood for grew in importance, the Muskogee oil operator became known as a bad man to keep as an enemy. Even his powerful adversary Robert Kerr realized this. Early in his 1960 reelection campaign Kerr contacted Oklahoma congressman Ed Edmondson and asked if he would discreetly inquire whether Simmons would be willing to bury the hatchet. The tables had been turned since the Liberia deal. This time it was Simmons who rebuked Kerr. "Jake became sought after as a man to give support," recalls Edmondson, who had several "discreet discussions" with Simmons on the subject. "I talked to him and Jake would not listen. He wasn't going to be for Kerr. I had to report back to Bob that I couldn't do anything."

Later in 1960 Simmons and Kerr did meet again. Although they never reconciled their differences, the two men found something they could agree on: Lyndon Johnson's candidacy for President. During the campaign, while Simmons was visiting Washington, Kerr introduced him to the Texan powerhouse. "He's always opposed me," Kerr reportedly told Johnson, "but I've always respected him." Johnson came to know and respect Simmons as well.[9]

True to Simmons's assessment during the 1960 primary, President Johnson delivered for African-Americans by successfully pushing through both the Civil Rights Act of 1964 and the Voting Rights Act of 1965. Black citizens who had never voted in their lives were being registered by the thousands. The debate over the necessity of equal opportunity for people of all races had, in most respectable quarters, shifted from a question of whether it was advisable to how it could best be implemented. A "War on Poverty" had begun. Public expenditures would check the economic crisis which created hopeless urban ghettos and squalid rural slums before the black underclass grew any larger. In it place would come "The Great Society," and African-Americans would be lifted with the rising ship. Money would placate the destructive rage which was beginning to explode in open conflict, like the August 1965 Watts riot, in which thirty-five people died and property damage surpassed $100 million.[10]

Jake Simmons had his own solution. The capitalist system he so enthusiastically embraced allowed all people a piece of the pie if they had a way to pay for it. Without jobs all a person could ever hope for were the crumbs which their government chose to give them. School desegregation, residential integration, even legislative equality were less important to him than securing jobs for blacks. Under welfare, Simmons believed, people grew lazy and despondent. Certainly government-bought food was better than no food at all, but it was no substitute for a decent opportunity. Simmons railed against every aspect of the system that limited black access to jobs. In the summer of 1967 he was appointed to a special "Employment Opportunity Committee" created by Muskogee's mayor. At the first meeting Simmons appealed to the city's employers to create jobs for blacks and train them. Mayor Jim Egan responded that it was "the general consensus of the committee that sufficiently trained Negroes were not available to fill Muskogee job vacancies when needed."[11]

Simmons did his best to challenge this claim. After losing his regular secretary during the early sixties he used the position in his office as a training ground for black women. He would hire an inexperienced person, then train her to answer phones and handle office machinery. Once she had acquired basic skills, he would lobby people he knew at banks or offices in town to interview her for job vacancies. If he was doing business with the office, he used that leverage as a pressure point, saying things like, "I've got deposits in your bank and you don't have any blacks working for you." His recommendation went a long way toward establishing credentials.[12]

Simmons also believed that the qualifications issue was often a cover for employment discrimination. When he came across an unintegrated workplace in Muskogee, he pulled whatever strings he could find to convince the business to hire black employees. If the establishment repeatedly turned down black applicants, he would unleash his secret weapon: Almetta Carter.

Carter was a white supremacist's worst nightmare. A well-trained ebony-skinned woman, strongly built, attractive and articulate, she packed a pistol in her purse and had no trouble letting people know when they were stepping on her toes. Carter, a Muskogee native, was in her early twenties when she went to work for Simmons in 1966. She had recently graduated from business college and was trained to type, take shorthand, do bookkeeping, and manage office machinery. Her first attempt to obtain employment through a local job agency resulted in an offer to be a chicken plucker. Jake Simmons tested Carter's skills, interviewed her, then hired her as a full-time secretary. She impressed him with her powerful backbone: a divorced single mother who refused to allow anyone to push her around.

Simmons decided that she was just the person to test the hiring policies of unintegrated businesses, like manufacturing companies or a large department store which had never hired a black salesperson. He would wait until they advertised for workers, then send his secretary to apply for a job. Carter would arrive, qualifications in hand. Out would come the written tests for prospective employees. Carter passed with flying colors. Next would come the personal interview—although in some respects it was really she who did the interviewing. "He sent me to have interviews with employers to see exactly how they were treating people," she recalls, "whether they were accepting people on their level of education and qualifications as opposed to their color. Then I would report back to him the same day."

In every instance but one, Carter, who probably was known to have had Simmons's backing, was given the job. She would work for a while at the new position, proving that business did not come to a standstill because of a black worker. Occasionally she would have to weather snide remarks and even attempts to sabotage her work, but she never had trouble standing up for herself. Once their point had been made, Simmons would help find another black person for the job, and Carter would return to his office. After a few years there were few large workplaces left in Muskogee to integrate.[13]

Jake Simmons became known to blacks across eastern Oklahoma as a man who would find employment for anyone willing to work, even if he had to call the governor to arrange it "He found jobs for hundreds of people," recalls Simmons's minister, the Reverend E. W. Dawkins. "People would just go to Jake's office as though it was an employment office. If he couldn't get to you today he'd say, 'Come back tomorrow, I'm going to call somebody and I'm going to get you a job.' And he would."

Every politician who knew Simmons—and there were few in Oklahoma who did not—learned to anticipate the requests which followed whenever he called. Jim Barker, currently the speaker of Oklahoma's House of Representatives and probably the most powerful legislator in the state, knew Simmons from his first days in office, as representative from Muskogee in 1968. Barker was impressed by the amount of time which Simmons spent helping people find jobs, especially poor people. He lobbied Barker on everything from jobs for highway laborers to legislation affecting specific work programs. Many times, Barker recalls, the phone in his Muskogee office would ring and he would hear a polite but determined voice on the other end.

"Representative," Simmons would begin, "can you go with me for a few minutes?"

"Where are we going, Mr. Simmons?" Barker would ask.

"I want you to meet someone," was Simmons's pat answer. "It is real important that you meet someone this afternoon."

"About the time I would say, 'Yes, I'll go, what time?' " Barker remembers, " he would say, 'I'll be by in ten minutes.' And really he would be out front in ten minutes and we would go see the individual."

It was never a power lunch with a well-heeled contributor. Simmons always took his representative to see an unemployed worker looking for a job, a student who wanted a college education, or a destitute mother confused about how to apply for public assistance for her newborn child. Barker, who still receives better than 90 percent of Muskogee's black vote, did his best to secure assistance for anyone Simmons brought to him for help. He says that Simmons never asked for anything personally. Far from being bothered by the activist's persistent solicitations. Barker thought they helped him do his job better, and appreciated Simmons for it.[14]

With a son in Washington Simmons asked a bit more from national politicians. Congressman Ed Edmondson remembers Simmons doggedly pressuring him to see that more blacks were appointed to the government-sponsored service academies at West Point and Annapolis. Locally he wanted more African-Americans hired for Muskogee's federally financed veterans' hospital and the U.S. courthouse. Edmondson also recalls Jake Simmons "pressing hard" for his son J.J. on several occasions. In the recommendation-dependent world of federal appointments, Jake pushed every button he could find to boost the prospects of J.J., who rose within the Department of Interior from domestic petroleum production specialist in 1962 to assistant director for the Office of Oil and Gas in 1966 to deputy administrator of the Oil Import Administration in 1968 to the influential administrator of the Oil Import Administration in 1969. Former Department of Commerce official Roy Gootenberg considered Jake "very much a political animal," who called him occasionally to help his son get through the bureaucracy of Washington's "old-boy network." Former congressman Jim Jones, also originally from Muskogee, had been a White House official in the Johnson administration since 1965 and served as chief of staff from late 1967 through 1968.[15] Now a Washington-based attorney, Jones regards Jake Simmons, Jr., as something of a "black Joe Kennedy in terms of pride and pushing his sons ahead of him." He says that President Johnson knew Simmons by name, and that "the impression from the White House was that they [the Simmonses] were never satisfied. We'd just get J.J. settled in one job and he was back there wanting another. Which is not necessarily a criticism. Ambition is something we all have in this town."[16]

By 1967 Jake Simmons's political stock in the capital was rising. When the head of Ghana's armed forces, Lieut. General J. N. Ankrah, was given a White House luncheon by President Lyndon Johnson in October, both Jake and Eva, as well as their Oklahoma senator Fred Harris, were invited to attend.[17]

Donald Simmons, who had introduced his father to the opportunities in Ghana, did not partake in the event. To his father's consternation, Donald was no longer participating in the family oil business. Swept up in the black power movement of the sixties, Donald moved to New York City and joined other successful black men his age trying to solve the problems of the urban ghetto. He was an angry young man whose consciousness had been raised by the activism of his contemporaries. Even while working for Phillips in Africa he had begun leaving his hair long, and regularly wore traditional African dress and beads as a statement of his ethnic heritage. When he returned from Nigeria he felt a pressing need to do his part, but sit-in demonstrations and other passive forms of protest went against his training, as did thc radicalism of revolutionary groups like the Black Panthers (whom he refers to as "proper b.s. artists whom an honest day's work would scare to death"). To Donald, doing his own thing for the people meant black economic development.

Simmons became the controversial executive director of the Harlem Commonwealth Council (HCC) in New York City. The council had been founded in 1967 with a $400,000 research and development grant to Columbia University from the federal Office of Economic Opportunity. Roy Innis ran the HCC for a year, with the intention of avoiding the city's "poverty bureaucracy" by creating an organization that would give black power an economic base. Innis left to join the Congress of Racial Equality (CORE) and Simmons took his place. He quickly became disgusted with the HCC's white economic advisers from Columbia University and the New School for Social Research. The advisory experts spent all but $47,000 of the federal funds on studies. Like his father, when Donald disagreed he disagreed loudly. He regarded the "experts" as "great white liberals" who were "ripping off money by the hundred thousands to turn out some graduate student thesis at the end of the year." In August 1969 *Business Week* reported the dispute between Donald's faction of black activists and the university theorists.[18]

The article noted that the experts believed that Harlem would never be able to stand on its own feet economically and therefore needed to rely on the development of low-yielding "greenhouse industries" which justified themselves not by profits but by the social benefits they produced. Simmons and his staff disdained this approach because it perpetuated a dependence on white funds. "The only way we can stop being a colony is to form profit-making enterprises," Donald insisted. "The academic economists are completely out of it," Donald told *Business Week*. "We have our own model for black economic development."

The magazine credited Simmons with bringing to the HCC "the independence that was always a necessary part of its mission." He raised $1 million in grants to start a local development company, which opened a sewing machine franchise and purchased the last foundry in Harlem—both intended to be independent profitable ventures. Simmons turned to management consultants, investment bankers, and law firms for specific advice, but refused to have anything to do with the group's original economists. He became one of the most sought-after executives in the field of independent black development. He helped run the Urban Coalition's Harlem operation, and spent a few days a week in Detroit managing a similar program for the Bank of Commerce. He also served as a dollar-a-year consultant for CORE, assisting with the development of new programs across the country.[19] Within three years, however, Donald grew disillusioned with the armchair bureaucracy involved in developing programs to help other people make money. He also struggled with a difficult divorce, and decided to leave New York before he became an unhappy urban workaholic. "I came to the realization that it may be creating a brave new world doing all these things for your race," he explains, "but it doesn't put any money in your pocket."

Which is not to suggest that Donald's family was hurting for money. By the time Donald rejoined his father's international royalty business, there was enough business to keep both of them occupied. By this time, according to his pastor E. W. Dawkins, Jake Simmons "stopped making change and started making money." He increased his philanthropic contributions to various groups, especially those related to the African Methodist Episcopal Church. There was Shorter College, the NAACP, the Boy Scouts, and a scholarship fund for Ward Chapel graduates. He gave more money to individuals, however, than groups, developing a reputation as a man who "didn't have no real limit" to what he would give a person in need. The line outside Simmons's inner office sometimes began in the waiting room and trickled over into the hallway. According to James Leake, a television station owner and Muskogee's wealthiest citizen, Simmons "always helped down-and-out people all over this country, regardless of their color." Like a Sicilian godfather Simmons received people's problems, whatever they were. By the sixties and seventies he was spending half his working hours working out other people's problems. Often his "consultations" would take him out of the office. For three decades mortgage banker Bill Friman, owner of Mercury Mortgages in Tulsa, was visited by Simmons at least once a month. Each time the oilman brought with him a black farmer or homeowner who needed a new mortgage. Friman never paid Simmons a cent for the business he brought him. "Jake knew a hell of a lot of people," Friman recalls. "If they needed money or had a problem they would see Jake. That was all there was to it. He was Kingfish of the blacks."[20]

Simmons loved giving advice, and was not at all averse to giving money. Usually the two went hand in hand. J.J. III notes that his father "was a man of the people—his favorite charity was the people. I mean the people who walked in and out of that office every day, and the people who called him day and night when they lost their jobs or when they had been discriminated against. And anybody in the world could get fifty dollars out of him with a hard-luck story. They just had to sit down and let him lecture them. If they were willing to go through that they could get anything they wanted to. That was worse than a whipping to have to go through that lecture."

The Reverend Dawkins remembers that when it came to hard-luck stories, Jake Simmons had a particularly soft spot for preachers. "Preachers were just preying on old Jake," Dawkins recalls. "One would go in there saying he was in bad shape, then the other. He put up the money to save the farm of one preacher, and [later] told me, 'It's not that he won't pay back the money; the fact is he won't even pay the interest.' "

The needy were not the only people Jake Simmons believed in giving money to. He also believed that because African-Americans were making demands on politicians, they should give contributions to ensure that their voices were heard. "He felt that blacks had a reputation of not contributing," recalls Ambassador Horace Dawson. "Therefore he thought it very important to contribute to the political process. He was a Democrat himself, but I think he had a technique of contributing to both Republican and Democrat, because he knew he had to deal with both and never wanted to be on the losing side." Simmons was especially assiduous in making donations to local Democrats. According to ten-term congressman Ed Edmondson, "He contributed to every campaign I was in. Anytime I had a fund-raiser he was there with a check."

Despite Jake Simmons's success he felt that many people underestimated him and would only accord him respect if he spoke up for himself. Wealth did not soften him; in this regard he remained a fighter until the day he died. A number of white oil executives that Simmons did business with expressed to me their admiration that he was able to disregard racism and carry on as though it did not exist. But this was simply the side of Jake Simmons he showed to whites. Disregarding race was fine in relating to other races, but to other African-Americans it was a luxury which the world around them would not allow. To blacks he spoke passionately of the need to be conscious of one's place in what he called "this pernicious system," to avoid being swept up like a pawn in another man's game. To know what was going on, he preached, one had to control conscientiously his own destiny. In a 1972 letter to his three sons, he admonished them to remember that "life of your own making is the sum total of your integrity." He believed many people sold themselves short, losing control of their fate because of a lack of discipline or a failure to set their goals high enough. "For a man not to achieve the things he has an opportunity to do is such a waste of life," he told J.J. III. "Not for selfishness, but for what you can do for others."[21]

To his friend Horace Dawson, the former U.S. ambassador to Botswana, Simmons was "in some ways a very angry man" who "felt automatically that prejudice was a fact of life and he was constantly battling it. He was combative. He felt that people would automatically slight him, automatically do him in, and he was ready to fight that at every turn."

Simmons was not afraid to fight. In fact, he came to believe that it was an important part of the development of the human spirit. In an inspirational essay he wrote late in his life, he noted, "I subscribe myself to the philosophy that mankind was made for struggle, for hardship, for enduring suffering. Such was the plight of our race, and who in the world will challenge our eighty years of progress in

America, which is unprecedented in the entire category of human endeavor? Mankind grows strong and noble under strain."[22]

Through his life Jake Simmons remained religiously adamant in his conviction that social justice was not an issue open to compromise. The Reverend E. W. Dawkins recalls one of the best examples of this attitude. In 1964 Dawkins accompanied Simmons as part of an elected five-person Muskogee delegation to the African Methodist Episcopal Church's important general conference, held once every four years to vote in new bishops and debate church policy. In 1964 more than a thousand delegates assembled at a large assembly hall in Cincinnati, Ohio. The civil rights movement had alarmed Cincinnati's authorities, and squads of armed local police ringed the hall, inside and out. The police department claimed it was for the delegates' own protection, but Simmons didn't buy it. He rushed to the stage, grabbed the microphone, and held up the entire conference, refusing to allow it to begin until the police vacated the premises. "This is a meeting of Christians," he yelled. "We can do our work without these police standing all over the place with their guns and blackjacks as though we were all criminals." After an hour-long standoff, the police department gave in and left. Simmons surrendered the microphone and the conference proceeded. He later paid the two-week hotel bill for Muskogee's entire delegation.[23]

Jake Simmons's confrontational energy did not dissipate with his youth. In 1969, during a highly charged meeting of a large community action program he helped run, Simmons argued angrily with a powerful white rival who was a fellow member of the board of directors about what he believed to be the racially motivated firing of two top staff members. The meeting was broken off early as Simmons, although near seventy, challenged his nemesis to "settle it outside" with fists.[24]

Notwithstanding his unique success in many areas, there were a few goals which Simmons was never able to attain. He would have liked a position as an ambassador to an African country, but despite his connections in Washington, such an appointment eluded him. Simmons fared no better when it came to elected office, although he wielded great regional power behind the scenes. In addition to his two unsuccessful bids for city council he made an effort, in late 1966, to win the Democratic nomination for the state house of representatives. Unable to assemble enough support to beat an incumbent, he withdrew before the election. In assessing why Simmons never went further in politics, Muskogee multimillionaire James Leake, who is an important player in the county's Democratic Party, notes, "I think that jealousy of Jake prevented him from being the [political] power figure he was capable of being." George Nigh, who worked closely with Simmons during his long tenure as governor and lieutenant governor of Oklahoma, observes that Simmons's disposition had something to do with his shortcomings as a candidate. "There are a lot of people who contribute greatly to government who could never get elected to public office," Nigh says. " He was a little to the point at times. I think politically Jake Simmons would have had trouble getting votes because he didn't mind telling you what he thought. There are those of us who don't mind telling you what we think but we don't—you also have to be political in some sense."[25]

Simmons was even unable to win the approval of important elements of Muskogee's business community. His failure had far more to do with lingering racism than Simmons's disposition. During the late sixties he badly wanted to become a member of the board of directors of one of his city's major banks. According to Tulsa banker Taft Welch, whose holding company then owned a 27 percent interest in the Commercial Bank & Trust Company [Muskogee's largest bank at the time], Simmons approached one of the directors in 1969 and offered to deposit a sum rumored to be $1 million into a checking account if he was made a member of the board. The offer was discussed at an unofficial closed-door meeting. Welch says that he was for the appointment, but members of the board from Muskogee threatened to quit if Simmons was made a director. "They would every one resign," Welch remembers. "They also said we'd probably lose half our deposits." Simmons was told there was no vacancy on the board. He took his money elsewhere.[26]

Only those closest to Simmons were aware of his unrealized ambitions, and even these people believed that his triumphs far outweighed his defeats. He used his success as a prop, a symbol of what other blacks could achieve if they set their minds to it. He was especially effective as a role model for children, whom he loved to talk to. More than anything he sought to inspire. Mrs. M. L. Sanders, one of Oklahoma's first NAACP officers, says, "Jake was a good role model, especially for young men. His achievement could show them what they could be. He wanted black people to become uplifted, to command and demand respect. He didn't have much patience with people who were in a slump and didn't want to pull out. He wanted them to step on out."

To groups of young people, Simmons would justify his meddlesome lectures with a timeworn anecdote about an old man commencing work on a new bridge. A youngster approaches and asks, "Why are you building this bridge? You're an old man and this is the close of the day."

"Young man," went the reply, "I am building this bridge for you."[27]

With adult audiences, Simmons sought to inspire by heralding his own accomplishments. "There wasn't a bashful bone in him at all," Jake's friend Doris Montgomery says. "Jake was domineering. He engulfed people and had no humility in him. Some black people in the state resented Jake because he would always tell about his success. I once heard a woman tell him, 'I don't like you because you're a braggart.' He answered like Dizzy Dean. He said, 'Madam, when you done done it you ain't bragging.' "

Like all people of strong character, it was Simmons's most outstanding qualities, particularly his assertiveness, which affected people's responses to him both positively and negatively. When it came to breaking new ground his "bulldog tenacity" was his greatest asset. He persuaded influential white business associates to recommend him for membership in numerous unintegrated organizations. He was the first black member of the state's Independent Oil Producers Organization, and broke the color barrier at Tulsa's prestigious Petroleum Club. He was never a clubby sort of person, but attended often enough to make his presence known. (He enjoyed telling the story about what happened soon after he joined when he walked out of the club after dinner with a visiting African official. The African, dressed in his colorful diplomat's uniform, was passing near the entranceway when a gruff Oklahoma oilman, thinking him a bellhop, tossed his coat at him to hang up.)[28]

Far more important than these local affiliations was Jake Simmons's 1969 appointment, by the U.S. Secretary of the Interior, to the

National Petroleum Council. The council had been formed in 1946 to advise the federal government on oil and gas policy. It was the oil industry's club of clubs. Membership was restricted to the CEOs of every multinational energy company in America and a few privileged independents. Simmons, the first black to become a member of the council, was repeatedly reappointed and served on the council for eleven years (his son Donald followed his tenure there). At the group's regular meetings Jake had the opportunity to introduce himself to those of the nation's top oil executives he did not already know—and see if they were interested in securing oil concessions in Africa.[29]

Simmons never slowed down. Through the seventies, the last decade of his life, he worked on oil deals around the world, winning concession licenses in Kenya and Trinidad, and pursuing prospects in Guyana, Gambia, Colombia, and even Australia. Few of these ventures panned out—which is not to say that he failed to make money on them. By this time Simmons was highly renowned in his field. Tulsan Jack Zarrow, who with his brother owned one of the country's largest oilfield supply and pipeline companies, invested with Simmons during the seventies. When Zarrow traveled with Jake to Senegal and Gambia to discuss concessions, they met with the presidents, not ministers, of both countries.[30]

At home in Oklahoma Simmons's prestige also continued to climb. He was appointed to the state's Tourism and Recreation Commission, where he immediately began pressing for the hiring of more blacks and the depiction of minorities and women in the agency's advertising. His political support became so important, says George Nigh, that senators, members of Congress, governors, and legislative leaders crowded him for both "political advice and good judgment." While Nigh was governor and lieutenant governor, he notes, "Every time I tried to do anything politically I wanted him on my side."

Simmons, according to Nigh, explained to him how blacks needed to have an equal opportunity for making decisions as well as finding employment. "I totally concur with that," Nigh says. "He set the foundation for major decisions that I made in state government." Nigh also recalls that Simmons, at this stage of his life, did not just press for the rights of blacks. "He was just as concerned about women and Indians and anybody else that was not being given opportunity.

I can hardly ever remember him saying, 'Black, black, black.' I more nearly remember him saying, 'Fairness, fairness, fairness.' "

To Jake Simmons, the greatest measure of personal success was not political influence or money. It was the success of his children. He never failed to talk about them, sometimes pretending to forget that he had recently told the listener about one of his son's latest achievements just so he could repeat it. Donald was the young prodigy he was training to follow in his international footsteps. J.J. III was the government powerhouse. Kenneth was the family scholar and Harvard-educated intellectual. Blanche was his only child who had not graduated from college, and he remained determined to do something about this. Even after she passed fifty, Simmons pestered his daughter to go back to school. In 1975, she did so. He helped support her for four years and paid her complete tuition to Nazareth College, where she earned two bachelor's degrees (in social work and behavioral science). Accompanying the periodic checks Blanche received from her father were urgent letters of encouragement to "My darling daughter Blanche." In one of them, which came with a $3,000 check, he wrote, "Remember, all your brothers are college and university graduates. So you must not fail the Simmons family's desire for our children and grand- and great-grandchildren."[31]

Simmons's children responded to their father's affection with reverence. During my initial interview with J.J. III, I started with some questions about his career, and his first response was "Let's get this straight. No matter what I achieve in my life, I'll never be one-tenth the man my father was." Perhaps the compliment Jake Simmons would most have appreciated is one which came from his business associate and close friend Henry Zarrow, who told me, "I never saw any children love their parents like those boys did."

Jake Simmons cherished his family, but he did not spend his final years sitting back watching them grow up. During the late sixties, while having a new home built, Jake and Eva moved into a segregated senior citizens' apartment complex and went stir-crazy. After a few short months Eva told her friend Doris Montgomery, "He wanted to integrate Cape Frank Manor. Well, he's integrated it. Now I want to get the hell out of here."

Instead of slowing down, Simmons kept doing business until he died of heart failure at the age of eighty in March 1981. (On the day of his funeral the county of Muskogee closed the courthouse, and dignitaries from as far as Ghana jammed into the funeral service at the Ward Chapel A.M.E. Church. Governor George Nigh delivered the eulogy.) Even in his late seventies, Jake Simmons carried under his arm a dog-eared manila folder containing a half-dozen potential deals, which he pulled out when he met with oil executives. Signal's W. H. Thompson recalls that Simmons "had a deal for every occasion," and repeatedly showed him maps of his farm in Haskell, insisting that there was still a lot of oil there to be discovered. (A few years after his death his son Donald drilled a few new wells there and did indeed find commercial quantities of oil.)[32]

As long as he lived Simmons believed that his last major deal, the Agri-Petco Salt Pond project in Ghana, would be his crowning legacy. Although virtually every other international oil company had left Ghana by the late 1970s, Simmons had single-handedly managed to convince one American company to send $55 million in much-needed capital to explore off the country's continental shelf. In 1978, as preliminary test results at Salt Pond signified Ghana's first commercial oil find, General F. W. K. Akuffo, the nation's head of state, announced plans to hold a grand "durbar" ceremony, to honor Jake Simmons and commemorate the formal "inauguration" of the Salt Pond oilfield.[33]

In late November, 1978, Jake Simmons came to Ghana as a special guest of the government. He brought Eva and his son Kenneth to attend the special ceremony. The family stayed at a government guest house in Accra. Early in the morning on November 24 a state car chauffeured them to a large open field near the beach at Salt Pond.

A grand durbar is a massive public event in Ghana, and this one was designed to coincide with an important tribal ceremony in the central region of the country. In addition to the diplomats and government officials who streamed in from Accra, the event attracted hordes of citizens from all over the country. "Big" chiefs, "middle-ranking" chiefs, and "subchiefs" arrived with clans and designated "hornblowers." Most of the audience traveled by bus, or crammed on to "trot-trots," open-back trucks which carry scores of standing riders. Michael Akotu-Sasu, who came with the oil ministry, estimated that as many as 200,000 Ghanaians attended the day-long ceremony. There was drumming, dancing, and the singing of both tribal and Christian religious songs. The government provided food for its foreign guests and officials, although most people were fed by their clans' chiefs and subchiefs.

General Akuffo came by helicopter. He greeted the important chiefs respectfully (in many districts they wield far greater power than the government), and joined them in prayer and speeches, telling the multitude that it was the "government's hopeful expectation that the modest beginning of Salt Pond would usher the country into an oil boom." Ghana's newspapers featured the event on their front pages. The *Daily Graphic* headlined its coverage A DREAM COMES TRUE, noting that the significance of the occasion was "that Ghana has joined the fraternity of oil-producing countries at a time she is beset with serious economic problems."

The Simmonses sat on the dignitaries' dais. Jake and Eva were called together to the speakers' platform and Ghana's head of state presented the Oklahoman with his country's Grand Medal. Akuffo praised Jake and Donald's eleven-year involvement in his country, making it clear that the millions of dollars brought in to develop Ghana's oil prospects might never have come without their participation. "These men," he said, "apart from encouraging many American companies to undertake oil prospecting in Ghana, assigned part of their oil concessions to Agri-Petco."

Jake Simmons would live for barely two years beyond the grand durbar at Salt Pond. (On his death bed, he expressed disbelief that the God who had served him so well was letting him down.) His survivors would receive royalties on what proved to be a multimillion-barrel field for much of the next decade, but not the kind of money they would have made had Jake Simmons been as lucky as he was persistent. The Agri-Petco wells would produce for a while, trickle down to six hundred barrels a day, and finally be abandoned as uneconomic during the mid-eighties. Phillips Petroleum also left Ghana, favoring its projects in the Ivory Coast. Only minor exploration has taken place since, and the billion-barrel find off the coast of Ghana, if it indeed exists, remains unfound.

Jake Simmons would never live to see all this, and would have scoffed at any suggestion of defeat. He told the Ghanaian people that they had every reason to hope for a "far-reaching" oil-exploration program in 1979. He had become a national hero, a bridge between two worlds and an example of international black synergy. On that sunny afternoon in November, with his proud wife standing next to him, hundreds of thousands of Africans cheered as Ghana's head of state pinned the country's highest medal to his chest. As he looked out over the production platform floating in the ocean, there was no question in his mind that millions and millions of barrels of oil waited beneath them, oil which would enrich both the Ghanaian people and his own family. This time he was sure his optimism would pay off. Life was full of promise, and the earth abounded in untapped riches.[34]

Acknowledgments

This book could not have been written without the help of Jake Simmons's sons Donald and J. J. III. Their input, both as sources of information and as conscientious readers, was indispensable. Commissioner J. J. Simmons III's candor helped furnish a critical compass, and Donald's consistent interest in raising my consciousness provided a much-needed intellectual bridge from my home in faraway New York.

During the year I spent in rural Oklahoma and my many additional visits, Donald and Barbara Simmons were so supportive and treated me so much like a member of the family that I almost forgot I was a Yankee. Donald's secretary, Almetta Carter, provided me with both logistical support and, with the help of her fun-loving family, ceaseless entertainment. Other friends and relatives of the Simmons family also gave me a warm welcome. Their sister Blanche Jamierson sheltered me in Battle Creek, Michigan, when I got snowed in, and woke up in the middle of the night to make me homeopathic tea when I found myself too ill to sleep. Erma Threat and her husband, David, made me welcome in their Oklahoma City home; and everybody's favorite Uncle Johnny, Jake Simmons, Jr.'s, brother (and the last surviving member of his generation), told such good stories that after a while I stopped caring whether they were true or not.

A number of other Oklahomans made me feel like part of a tremendously nurturing community. Council Oak Books' delightfully zany publisher Paulette Millichap and the *Tulsa Tribune*'s crack investigative reporting chief Mary Hargrove were both great friends and allies. Bill and Ressie Brooks were helpful neighbors, made excellent cornbread, and taught me what little I know about farming.

The libraries in Oklahoma are a wonderful resource for a writer. Few states take more pride in their history, and Oklahoma's librarians and archivists show it. Bill Welge at the Oklahoma Historical Society was extremely helpful, as was his boss Mary Lee Boyle. At the University of Oklahoma's Western History Collection I was lucky enough to find Jack Haley just a few days before he retired, and also received assistance from Brad Garnard, Ray Miles, and Kelly Ross. Oklahoma City Library researcher Mary Patton copied clips for me which I could not locate, and Pat Doen in Okmulgee's tiny library was also handy with microfilm. At the Gilcrease Museum in Tulsa, librarian Sarah Erwin guided me through the rare manuscript collection. Tulsa University's archivist of special collections, Sid Huttner, was also a helpful guide.

At Tuskegee University, registrar Dorothy B. Conley was able to recover seventy-year-old school records for me; and at the National Archives in Washington, Bob Kvasnicka and William E. Lind copied and sent me other important documents.

In the private sector Jim and Ed Goodwin, Jr., the owners of Tulsa's black newspaper, the *Oklahoma Eagle,* allowed me to spend days sifting through and copying crumbling old editions, despite the fact that their father and Jake Simmons didn't always see eye to eye. Phillips Petroleum public relations man Bill Adams did his best to make me feel at home and helped me find my way to both the right people and papers. Forrest Brokaw at Sun Oil and Donna McFarland at Kerr-McGee were also helpful, as was former Sun public relations chief Chuck Snacke.

In Ghana Fred Tamakloe, steward at the Flair guest house, helped me learn the ropes. Sylvan Amegashie and his London associate

Tommy Thomas assisted with travel plans, and Sylvan's son Kojo was very hospitable once I got to Accra.

At various times over the past six years I hired different people to provide clerical and research assistance. Jake Simmons's great nephew Ahmed Shadeed spent month scrutinizing his family's history with me, and also left me with a greater understanding of the Islamic religion. Pat Thompson and Laura Richardson helped transcribe and type a mountain of taped interviews. Dear friends Pamela Korp and Carolyn McDonald helped with both basic research and moral support. My mother, Jane Greer, wielded an editing pencil with surprising dexterity, and she didn't even charge for it. Jay Gissen's further ability to reshape the manuscript at an unwieldy stage was invaluable, as he was one of the few colleagues I knew who could argue successfully for all the cuts that were needed and live to tell the tale. In the final weeks Anne Harley assisted with the bibliography and footnotes. I wish to thank all these helpers, and ask their forgiveness for my compulsive workaholic demands.

Finally I would like to thank my allies in the publishing world. First and foremost is my agent Lizzie Grossman, at Sterling Lord Literistic, who stood by me year after year, offering encouragement at critical moments and providing assistance which went far beyond the call of an agent's normal duty. At Atheneum I appreciate the groundwork which both Ann Rittenberg and Susan Leon did for the book, and am particularly grateful to my editor Erika Goldman for guiding me through the book's long overdue completion with a smile on my face.

Bibliography

Selected Books

Abel, Anne Heloise. *The American Indian as Participant in the Civil War.* Cleveland, Ohio: Arthur H. Clark Company, 1919.

———. *The American Indian as Slaveholder and Secessionist.* Cleveland, Ohio: Arthur H. Clark Company, 1915.

———. *The American Indian Under Reconstruction.* Cleveland, Ohio: Arthur H. Clark Company, 1925.

Achebe, Chinua. *Things Fall Apart.* New York: Fawcett Crest, 1960.

Adoff, Arnold, ed. *Black on Black.* New York: Macmillan, 1968.

Ahearn, Robert G. *In Search of Canaan.* Lawrence, Kansas: Regents Press of Kansas, 1978.

Aldrich, Gene. *Black Heritage of Oklahoma.* Edmond, Oklahoma: Thompson Book and Supply Co., 1973.

Allen, Philip M., and Aaron Segal. *The Traveler's Africa: A Guide to the Entire Continent.* New York: Hopkinson and Black, 1973.

Bailey, M. Thomas. *Reconstruction in Indian Territory.* Port Washington, New York: Kennikat Press, 1972.

Beadle, John Hanson. *The Underdeveloped West; or Five Years in the Territories.* Philadelphia: National Publishing Co., 1873.

Bennett, Lerone, Jr. *The Shaping of Black America.* Chicago: Johnson Publishing, 1975.

———. *Before the Mayflower.* 4th ed. Chicago: Johnson Publishing, 1969.

Bond, Horace Mann. *Education for Freedom.* Pennsylvania: Lincoln University, 1976.

Butler, Cleora. *Cleora's Kitchen.* Tulsa, Oklahoma: Council Oak Books, 1985.

Campbell, John Bert. *Campbell's Abstract of Creek Freedman Census Cards and Index*. Muskogee, Oklahoma: Oklahoma Printing Co., 1915.

Crockett, Norman L. *The Black Towns*. Lawrence, Kansas: Regents Press of Kansas, 1979.

Daily Times. *Nigeria Year Book: 1966; 1967*. Lagos, Nigeria: Time Press Limited, 1966, 1967.

Debo, Angie. *And Still the Waters Run: The Betrayal of the Five Civilized Tribes*. New York: Alfred A. Knopf, 1944.

———. *Prairie City*. Tulsa, Oklahoma: Council Oak Books, 1985.

———. *The Road to Disappearance: A History of the Creek Indians*. Norman, Oklahoma: University of Oklahoma Press, 1941.

Du Bois, W. E. B. *Black Reconstruction in America*. New York: Harcourt, Brace, 1935.

———. *The Souls of Black Folk*. Chicago: A. L. McClurg & Co., 1903.

Durham, Philip, and Everett L. Jones. *The Negro Cowboys*. New York: Dodd, Mead and Company, 1965.

Ellison, Ralph. *Invisible Man*. New York: Random House, 1952.

Ellsworth, Scott. *Death in a Promised Land: The Tulsa Race Riot of 1921*. Baton Rouge, Louisiana: Louisiana State University Press, 1982.

Epple, Jess C. *Honey Springs Depot*. Muskogee, Oklahoma: Hoffman Printing Corporation, 1964.

Fast, Howard. *Freedom Road*. New York: Crown, 1944.

Fisher, Leroy H., ed. *Oklahoma's Governors, 1929–1955; Depression to Prosperity*. Oklahoma City, Oklahoma: Oklahoma Historical Society, 1983.

Foreman, Grant, ed. *A Traveler in Indian Territory*. Cedar Rapids, Iowa: The Torch Press, 1930.

———. *Indian Removal*. Norman, Oklahoma: University of Oklahoma Press, 1932.

———. *The Five Civilized Tribes*. Norman, Oklahoma: University of Oklahoma Press, 1934.

Franklin, Jimmie Lewis. *Journey Toward Hope: A History of Blacks in Oklahoma*. Norman, Oklahoma: University of Oklahoma Press, 1982.

Franklin, John Hope. *From Slavery to Freedom: A History of Negro Americans*. New York: Alfred A. Knopf, 1947.

Freedom Fighters Edition. *Axioms of Kwame Nkrumah*. London: Panaf Books, 1969.

Garvey, Marcus. *Philosophy & Opinions of Marcus Garvey*. Vols. 1 and 2. New York: Atheneum, 1986.

Gibson, Arrell M. *The Oklahoma Story*. Norman, Oklahoma: University of Oklahoma Press, 1978.

Giddings, Joshua R. *Exiles of Florida*. Columbus, Ohio: Follett, Foster and Co., 1858.

Gideon, D. C. *Indian Territory: Descriptive, Biographical and Genealogical, including the Landed Estates, Country Seats, etc., etc. with a General History of the Territory*. New York: Lewis Publishing Co., 1901.

Green, Michael D. *The Creeks: A Critical Bibliography*. Bloomington, Indiana and London: Indiana University Press, 1979.

Gregory, Robert, *Oil in Oklahoma*. Muskogee, Oklahoma: Leake Industries, Inc., 1976.

Griffin, John Howard. *Black Like Me*. Boston: Houghton Mifflin, 1961.

Guthrie, Woody. *Bound for Glory*. New York: E. P. Dutton, 1943.

Haley, Alex, and Malcolm X. *The Autobiography of Malcolm X*. New York: Grove Press, 1964.

Haley, J. Evetts. *A Texan Looks at Lyndon*. Canyon, Texas: Palo Duro Press, 1964.

Halliburton, R., Jr. *The Tulsa Race War of 1921*. San Francisco: Rand E. Research Associates, 1975.

Harlan, Louis R. *Booker T. Washington: The Wizard of Tuskegee, 1901–1915*. New York and Oxford: Oxford University Press, 1983.

Harlan, Louis R., and Raymond W. Smock, eds. *The Booker T. Washington Papers*. Vol. 12 (1912–14). Urbana, Chicago, and London: University of Illinois Press, 1982.

Hawkins, Benjamin. *A Sketch of the Creek Country in the Years 1798 and 1799*. Savannah: Georgia Historical Society, 1848.

Hawkins, Hugh. *Booker T. Washington and His Critics*. Lexington, Massachusetts: D. C. Heath & Co., 1974.

Hofstadter, Richard, ed. *The United States 1916–1970: A World Power*. 4th ed. Englewood Cliffs, New Jersey: Prentice-Hall, 1976.

Hughes, Langston. *Fight for Freedom: The Story of the NAACP*. New York: Berkley Publishing, 1962.

Kaplan, Irving, et al. *Area Handbook for Ghana*. Washington, D.C.: The American University, 1971.

Katz, William Loren. *Black People Who Made the Old West*. New York: Thomas Y. Crowell Company, 1977.

———. *Black Indians*. Garden City, New York: Atheneum, 1986.

———. *The Black West*. Garden City, New York: Doubleday, 1971.

Kearns, Doris. *Lyndon Johnson and the American Dream*. New York: Harper & Row, 1976.

Kingston, Mike, ed. *Texas Almanac 1988–1989*. Dallas: Dallas Morning News, 1987.

Klein, Joe. *Woody Guthrie: A Life*. New York: Alfred A. Knopf, 1980.

Knowles, Ruth Sheldon. *The Greatest Gamblers*. Norman, Oklahoma: University of Oklahoma Press, 1978.

Lacey, Robert. *Ford: The Men and the Machine*. Boston and Toronto: Little, Brown, 1986.

Leckie, William H. *The Buffalo Soldiers*. Norman, Oklahoma: University of Oklahoma Press, 1979.

Lewis, David L. *King: A Biography*. Urbana, Illinois: University of Illinois Press, 1978.

Liberia: Geographical Mosaics of the Land and the People. Monrovia, Liberia: Liberian Ministry of Information, Cultural Affairs, and Tourism, 1979.

Littlefield, Daniel F. *Africans and Creeks: From the Colonial Period to the Civil War*. Contributions in Afro-American and African Studies, no. 47. Westport, Connecticut, and London: Greenwood Press, 1979.

Loomis, Augustus Ward. *Scenes in the Indian Country*. Philadelphia: Presbyterian Board of Publication, 1859.

Luper, Clara. *Behold the Walls*. Oklahoma City, Oklahoma: Wire Publishing, 1979.

Mahon, John K. *History of the Second Seminole War*. Gainesville, Florida: University of Florida Press, 1967.

McKenney, Thomas L. *Memoirs, Official and Personal: Thomas L. McKenney*. Lincoln, Nebraska: University of Nebraska Press, 1973.

McReynolds, Edwin C. *Oklahoma: A History of the Sooner State*. Norman, Oklahoma: University of Oklahoma Press, 1954.

Murray, William Henry. *The Speeches of William Henry Murray*, ed. A. L. Beckett and Victor E. Harlow. Oklahoma City: Harlow Publishing, 1931.

———. *The Negro's Place in Call of Race*. Tishomingo, Oklahoma: William H. Murray, 1948.

Nelson, Harold D., ed. *Nigeria: A Country Study*. Washington, D.C.: The American University, 1982.

Oklahoma Department of Libraries. *Directory of Oklahoma*. Oklahoma City: Oklahoma Department of Libraries, 1985–86.

Okmulgee Historical Society and Heritage Society of America, eds. *History of Okmulgee County, Oklahoma*. Tulsa, Oklahoma: Historical Enterprises, 1985.

Oquaye, Mike. *Politics in Ghana, 1972–1979*. Accra, Ghana: Tornado Publishers, 1980.

Painter, Neil Irwin. *Exodusters: Black Migration to Kansas After Reconstruction*. New York: Alfred A. Knopf, 1976.

Porter, Kenneth Wiggins. *The Negro on the American Frontier*. New York: Arno Press and the *New York Times*, 1971.

Quarles, Benjamin. *The Negro in the Making of America*. New York: Collier, 1971.

Rawick, George P., ed. *The American Slave: A Composite Autobiography*. Vols. 7 and 12. Westport, Connecticut: Greenwood Press, 1977.

Rister, Carl Coke. *Oil! Titan of the Southwest*. Norman, Oklahoma: University of Oklahoma Press, 1949.

Roberts, Thomas D., et al. *Area Handbook for Liberia*. Washington, D.C.: The American University, 1972.

Ross, B. Joyce. *J. E. Spingarn and the Rise of the NAACP*. New York: Atheneum, 1972.

Sampson, Anthony. *The Seven Sisters*. New York: Viking, 1975.

Smith, Robert A. *The Liberia Annual Economic Review*. Monrovia, Liberia: Providence Publications; New York: Liberia Investment Services Corporation, 1972.

Sprague, John T. *The Origin, Progress and Conclusion of the Florida War*. New York: D. Appleton and Co., 1848.

Stokes, Anson Phelps. *Tuskegee Institute: The First Fifty Years*. Tuskegee, Alabama: Tuskegee Institute Press, 1931.

Teall, Kaye M., ed. *Black History in Oklahoma: A Resource Book*. Oklahoma City, Oklahoma: Oklahoma City Public Schools, 1971.

Thirty Years of Lynching in the United States. New York: National Association for the Advancement of Colored People, 1919.

Thoburn, Joseph B., and Muriel H. Wright. *Oklahoma: A History of the State and Its People*. 4 vols. New York: Lewis Historical Publishing Co., 1929.

Thrasher, Max Bennett. *Tuskegee: Its Story and Its Work*. New York: Negro Universities Press, 1969.

Thwaites, Reuben Gold, ed. *Early Western Travels 1748–1846*. Cleveland, Ohio: The Arthur H. Clark Company, 1906.

Tolson, Arthur L. *The Black Oklahomans, A History: 1541–1972*. New Orleans: Edwards Printing, 1974.

Washington, Booker T. *Up from Slavery*. New York: Doubleday, 1901.

Washington, Nathaniel J. *Historical Development of the Negro in Oklahoma*. Tulsa, Oklahoma: Dexter Publishing Company, 1948.

Wertz, William C., ed. *Phillips—The First 66 Years*. Bartlesville, Oklahoma: Phillips Petroleum Company, 1983.

Woodson, Carter G. *A Century of Negro Migration*. New York: Russell and Russell, 1969.

Woodward, Thomas S. *Woodward's Reminiscences*. Montgomery, Alabama: Barrett and Wimbish, 1859.

Writers' Program of Work Projects Administration in the State of Alabama, ed. *Alabama: A Guide to the Deep South*. New York: Richard R. Smith, 1941.

Yankah, Kojo. *The Trial of J. J. Rawlings*. Tema, Ghana: Ghana Publishing Corporation, 1986.

Selected Magazine and Journal Articles

Alvarez, A. "Offshore," *New Yorker*, January 20 and 27, 1986.

Balman, Gail. "The Creek Treaty of 1866," *Chronicles of Oklahoma*, Summer 1970.

Banks, Dean. "Civil-War Refugees from Indian Territory in the North, 1861–1864," *Chronicles of Oklahoma*, Autumn 1963.

Bennett, Lerone, Jr. "Chronicles of Black Courage," *Ebony*, September 1983.

Bittle, William E., and Gilbert L. Geis. "Racial Self-Fulfillment and the Rise of an All-Negro Community in Oklahoma," *Phylon*, July 1957.

Boyd, Mark F. "The Seminole War: Its Background and Onset," *Florida Historical Quarterly*, July 1951.

Carter, Tom. "James Leake," *OK Magazine*, July 6, 1986.

Chapman, Berlin B. "Freedmen and the Oklahoma Lands," *The Southwestern Social Science Quarterly*, September 1948.

Childs, Thomas. "Extracts from His Correspondence with His Family," *Historical Magazine*, 3rd ser., March 1874, April 1875.

Douglass, Frederick. "Remarks on This Exodus by Frederick Douglass," *Journal of Negro History*, January 1919.

DuBois, Shirley Graham. "What Happened in Ghana: The Inside Story," *Freedomways*, Summer 1964.

Fathree, Cheryl. "All Black State Was the Goal," *OK Magazine*, June 22, 1986.

Fleming, Walter L. " 'Pap' Singleton was the Moses of the Colored Exodus," *American Journal of Sociology*, July 1909.

Foreman, Carolyn Thomas. "Alexander McGillivray, Emperor of the Creek," *Chronicles of Oklahoma*, Spring 1929.

———. "The Light-Horse in the Indian Territory," *Chronicles of Oklahoma*, Spring 1956.

Forry, Samuel. "Letters of Samuel Forry, Surgeon, U.S. Army, 1837–1838," *Florida Historical Quarterly*, January 1928.

Franks, Kenny A. "The Confederate States and the Five Civilized Tribes: A Breakdown of Relations," *Journal of the West,* July 1973.

Garvin, Roy. "Benjamin, or 'Pap' Singleton and His Followers," *Journal of Negro History,* January 1948.

"Ghana May Join Offshore Oil Producers," *The Oil and Gas Journal,* July 13, 1970.

Graham, Richard C. "Nigeria: Oil Country Now," *International Commerce,* July 20, 1964.

Gray, Seymour. "Investing in Ghana," *Black Enterprise,* July 1977.

Harmon, Bill. "When Men Forget Color," *Oklahoma's Orbit,* August 25, 1963.

Hill, Mozell C. "The All-Negro Communities of Oklahoma: The Natural History of a Social Movement," *Journal of Negro History,* July 1945.

———. "Basic Racial Attitudes Towards Whites in the Oklahoma All-Negro Community," *American Journal of Sociology,* May 1944.

———. "A Comparative Study of Race Attitudes in the American Negro Community in Oklahoma," *Phylon,* July 1946.

"Honoring 'President' Tubman," *The Nation,* January 26, 1952.

Jaffe, Thomas. "Don't Fill Up on Phillips," *Forbes,* November 17, 1986.

Johnston, J. H. "Documentary Evidence of the Relations of Negroes and Indians," *Journal of Negro History,* Winter 1929.

Kraft, Joseph. " 'King' of the U.S. Senate," *Saturday Evening Post,* January 11, 1963.

Kremer, Gary R. "For Justice and a Fee: James Milton Turner and the Cherokee Freedmen," *Chronicles of Oklahoma,* Winter 1980–81.

Lamberts, O. A. "Historical Sketch of Col. Samuel Checote, Once Chief of the Creek Nation," *Chronicles of Oklahoma,* Autumn 1926.

Lauderdale, Virginia E. "Tullahassee Mission," *Chronicles of Oklahoma,* Autumn 1948.

Liberian Embassy, Washington, D.C. *Liberia Today,* July 1952.

Massaquoi, Hans J. "Investment Opportunities in Nigeria," *Ebony,* March 1964.

Mellinger, Paul. "Discrimination and Statehood in Oklahoma," *Chronicles of Oklahoma,* Autumn 1971.

Meserve, John Bartlett. "Chief Opothleayahola," *Chronicles of Oklahoma,* Vol. 9, Winter 1931–32.

———. "Chief Pleasant Porter," *Chronicles of Oklahoma,* Vol. 9, Autumn 1931.

Miller, Worth Robert. "Frontier Politics: The Bases of Partisan Choice in Oklahoma Territory, 1890–1904," *Chronicles of Oklahoma,* Winter 1984–85.

Morton, Ohland. "The Government of the Creek Indians," *Chronicles of Oklahoma,* Spring 1930.

Muwakkil, Salim. "Marcus Garvey: Black Moses Celebrated by Aged Followers," *In These Times,* September 10–16, 1986.

Neuringer, Sheldon. "Governor Walton's War on the Ku Klux Klan: An Episode in Oklahoma History, 1923–1924," *Chronicles of Oklahoma,* Summer 1967.

Nicholson, Mark, and Setorwu Gagakuma. "Ghana: A New African Survey," *New African,* January 1986.

Nkrumah, Kwame. "The Mechanisms of Neo-Colonialism," *Freedomways,* Spring 1966.

Norman, James R. "What the Raiders Did to Phillips Petroleum," *Business Week,* March 17, 1986.

Peterson, Sarah. "Bartlesville, a 'Company Town' and Proud of It," *U.S. News & World Report,* October 22, 1984.

Phelps, John W. "Letters of Lieutenant John W. Phelps, U.S.A., 1837–1838," *Florida Historical Quarterly,* October 1927.

"Phillips Drills Hefty Strike off Ivory Coast," *The Oil and Gas Journal,* February 21, 1983.

Porter, Kenneth Wiggins. "The Founder of the 'Seminole Nation' Secoffee or Cowkeeper," *The Florida Historical Quarterly,* April 1949.

———. "Florida Slaves and Free Negroes in the Seminole War, 1835–1842," *Journal of Negro History,* April 1943.

———. "Negro Guides and Interpreters in the Early Stages of the Seminole War, December 28, 1835–March 6, 1837," *Journal of Negro History,* April 1950.

———. "Negroes and the East Florida Annexation Plot, 1811–1813," *Journal of Negro History,* January 1945.

———. "Negroes and the Seminole War, 1835–1842," *Journal of Southern History,* November 1964.

Rampp, Lary C. "Negro Troop Activity in Indian Territory, 1863–1865," *Chronicles of Oklahoma,* Spring 1969.

Roberson, Jane W. "Edward P. McCabe and the Langston Experiment," *Chronicles of Oklahoma,* Autumn 1973.

Savage, Sherman W. "The Negro in the Westward Movement," *Journal of Negro History,* October 1940.

Savage, William W., Jr. "Creek Colonization in Oklahoma," *Chronicles of Oklahoma,* Spring 1976.

Seligman, Daniel. "Senator Bob Kerr, the Oklahoma Gusher," *Fortune,* March 1959.

"Senator Kerr Talks about Conflict of Interest," *U.S. News & World Report*, September 3, 1962.
"Signal Combines Slates Wildcat off Ghana," *The Oil and Gas Journal*, December 15, 1969.
Simmons, Jake. "From Cowboy to Landlord," *Southern Workman*, March 1915.
Smith, Patrick. "Confessions of the CIA," *New African*, December 1985.
Sullivant, Otis. "Rich Man's Race," *The Nation*, June 26, 1954.
Turner, C. W. "Events Among the Muskogees During Sixty Years," *Chronicles of Oklahoma*, Spring 1932.
Valliere, Kenneth L. "The Creek War of 1836, a Military History," *Chronicles of Oklahoma*, Winter 1979–80.
Whitaker, Charles, "W. E. B. Du Bois: A Final Resting Place for an Afro-American Giant," *Ebony*, November 1986.
Williams, Nudie E. "Black Men Who Wore the Star," *Chronicles of Oklahoma*, Spring 1981.
———. "The Black Press in Oklahoma: The Formative Years, 1889–1907," *Chronicles of Oklahoma*, Autumn 1983.
Willson, Walt. "Freedmen in Indian Territory During Reconstruction," *Chronicles of Oklahoma*, Summer 1971.
Windom, William. "The Senate Report on the Exodus of 1879," *Journal of Negro History*, January 1919.

Selected Government Documents

American State Papers: V. Military Affairs. Vols. 6 and 7.
Bureau of American Ethnology, Smithsonian Institution. *Early History of the Creek Indians and Their Neighbors*. Prepared by John R. Swanton. Bulletin 73. Washington, D.C.: U.S. Government Printing Office, 1922.
Congressional Globe. 46 vols. Washington, D.C., 1834–73.
Congressional Record. Washington D.C., 1873– .
National Archives. Records of the Office of the Adjutant General. Record Group 94. General Jessup's Papers.
———. Records of the Office of Indian Affairs. Record Group 18. Register of Letters Received, 1824–1880.
———. Records of the Office of Indian Affairs. Record Group 75. Letters from Creek Agency to the Committee of Indian Affairs.

Senate Committee on Indian Affairs. 1904. *Laws and Treaties,* Vol. 2. Charles J. Kappler, ed. Washington, D.C.: U.S. Government Printing Office.

U.S. Bureau of the Census. Part 1 of *Census of the United States: 1900, Agriculture,* Vol. 5. Washington, D.C.: U.S. Government Printing Office, 1902.

———. *Census of the United States: 1910, Agriculture,* Vol. 7. Washington, D.C.: U.S. Government Printing Office, 1913.

U.S. Bureau of the Census. *Indian Lands West of Arkansas.* (*Oklahoma*) *Population Schedule of the United States Census of 1860.* Transcribed by Frances Woods, 1964. Oklahoma City, Oklahoma: Oklahoma Historical Society.

———. Senate. 25th Congress. Ex. Doc. 78. Serial 323.

———. Senate. 25th Congress. Ex. Doc. 225. Serial 348.

———. Senate. Committee on Indian Affairs. *Letter of the Secretary of the Interior.* 40th Congress, 2nd session, 1868. Ex. Doc. 64.

———. House. *Letter from the Secretary of the Interior.* 41st Congress. 2nd session, 1870. Ex. Doc. 217.

———. House. Committee on Indian Affairs. *Petition of the Delegates of the Creek Nation.* 45th Congress, 2nd session, 1878. Misc. Doc. 38.

———. Senate. Committee on Indian Affairs. *Claims of the Loyal Creeks.* 57th Congress, 1st session, 1902. S. Rept. 420.

U.S. Department of Commerce. Part 1 of *Historical Statistics of the United States: Colonial Times to 1970.* Washington, D.C.: U.S. Government Printing Office, 1975.

U.S. Department of the Interior. *Report of the Commissioner of Indian Affairs to the Secretary of the Interior.* Washington, D.C.: U.S. Government Printing Office, 1861, 1862, 1863, 1864, 1865, 1866, 1867, 1868, 1869, 1870, 1871.

U.S. Department of War. *War of Rebellion: Official Records of the Union and Confederate Armies.* 1st series, Vols. 34, 41, and 48. Washington, D.C.: U.S. Department of War, 1888–98.

Newspapers and Magazines

Black Dispatch (Oklahoma City)

Black Enterprise

Business Week

Congressional Record
Daily Graphic (Accra, Ghana)
Daily Oklahoman
Forbes
Fortune
Ghanaian Times (Accra, Ghana)
International Commerce
Muskogee Cimeter
Muskogee Daily Phoenix
Muskogee Indian Journal
Muskogee Times-Democrat
Nation
New Africa
New Republic
New York Times
Newsweek
Oil and Gas Journal
Oklahoma Daily Democrat
Oklahoma Eagle
Oklahoma Independent
Oklahoman Orbit
Okmulgee Chieftain
Tulsa Eagle
Tulsa Star
Tulsa Tribune
Tulsa World
U.S. News & World Report
World Oil

Unpublished Materials

Bartlesville Chamber of Commerce. *Bartlesville,* promotional brochure, 1986.

Gilcrease Museum. Creek Papers 1782–1893. Tulsa, Oklahoma.

Hickman, Gerald. "Disenfranchisement in Oklahoma: The Grandfather Clause of 1910–16." M.A. thesis, University of Tulsa, 1967.

Jackson, Nellie B. "Political and Economic History of the Negro in Indian Territory." M.A. thesis, University of Oklahoma, 1960.

Kerr-McGee. *1965 Annual Report*. Oklahoma City, Oklahoma.

Library of Congress. Manuscript Division. Booker T. Washington Papers.

———. Papers of the NAACP. Groups 1, 2, and 3.

Mulroy, Kevin. "Relations Between Blacks and Seminoles After Removal." Ph.D. diss., University of Keele, 1984.

Oklahoma Historical Society. Creek Tribal Records. Oklahoma City.

———. Grant Foreman Collection. Oklahoma City.

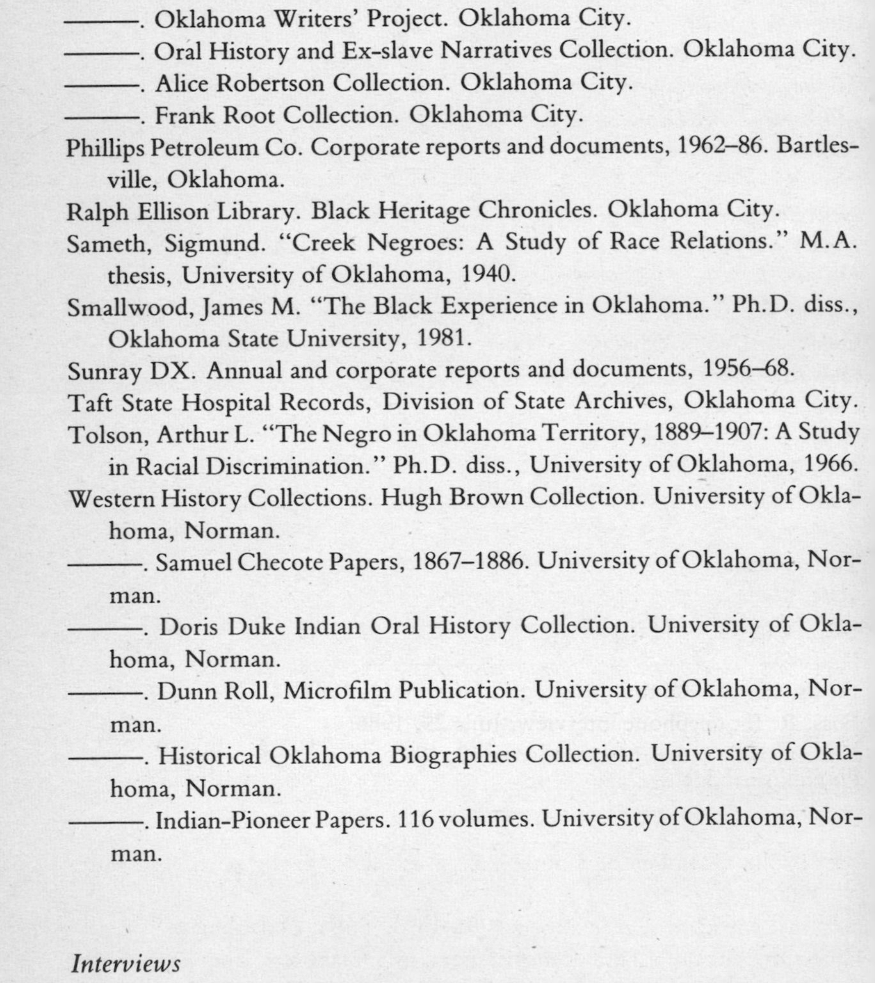

———. Oklahoma Writers' Project. Oklahoma City.

———. Oral History and Ex-slave Narratives Collection. Oklahoma City.

———. Alice Robertson Collection. Oklahoma City.

———. Frank Root Collection. Oklahoma City.

Phillips Petroleum Co. Corporate reports and documents, 1962–86. Bartlesville, Oklahoma.

Ralph Ellison Library. Black Heritage Chronicles. Oklahoma City.

Sameth, Sigmund. "Creek Negroes: A Study of Race Relations." M.A. thesis, University of Oklahoma, 1940.

Smallwood, James M. "The Black Experience in Oklahoma." Ph.D. diss., Oklahoma State University, 1981.

Sunray DX. Annual and corporate reports and documents, 1956–68.

Taft State Hospital Records, Division of State Archives, Oklahoma City.

Tolson, Arthur L. "The Negro in Oklahoma Territory, 1889–1907: A Study in Racial Discrimination." Ph.D. diss., University of Oklahoma, 1966.

Western History Collections. Hugh Brown Collection. University of Oklahoma, Norman.

———. Samuel Checote Papers, 1867–1886. University of Oklahoma, Norman.

———. Doris Duke Indian Oral History Collection. University of Oklahoma, Norman.

———. Dunn Roll, Microfilm Publication. University of Oklahoma, Norman.

———. Historical Oklahoma Biographies Collection. University of Oklahoma, Norman.

———. Indian-Pioneer Papers. 116 volumes. University of Oklahoma, Norman.

Interviews

(All interviews conducted by author.)

Adams, Bill: Bartlesville, Oklahoma, June 23, 1986.

Alexander, Zvi: New York, August 19, 1986.
Amegashie, Sylvan: London, August 3, 6, 1986.
Akotu-Sasu, Michael: Accra, Ghana, July 23, 1986.
Atkins, Hannah: telephone interview, June 3, 1986.
Barker, State Senator James: Oklahoma City, Oklahoma, March 11, 1986.
Barnes, Espanola: Muskogee, Oklahoma, December 9, 1984, and April 19, 1986.
Barnor, Dr. Matthew Anum: Accra, Ghana, July 20, 1986.
Bateman, Becky: telephone interview, June 17, 1986.
Billet, Marvin: New York: September 22, 1986.
Brokaw, Forrest: telephone interview, May 30, 1986.
Brown, Erma: Muskogee, September 28, 1983.
Brown, Gleaton: Haskell, Oklahoma, December 11, 1986.
Brown, Ophelia: Haskell, June 17, December 7, 1986.
Brown, Sam: Haskell, December 11, 1986.
Butler, Cleora: Muskogee, August 2, 1982; April 9, 1983.
Carter, Almetta: Muskogee, November 25, December 3, 1986.
Chancey, Elizabeth: telephone interview, March 6, 1986.
Chandler, Jesse, Dr.: Muskogee, June 20, 1986.
Chandler, Wayne: Oklahoma City, June 2, 1986.
Davis, Napoleon: Muskogee, December 15, 1986.
Dawkins, The Reverend E. W.: Tulsa, May 27, 1986.
Dawson, Ambassador Horace G., Jr.: Washington, D.C., September 24, 1986.
Dunlop, Rufus: Haskell, December 15, 1986.
Dennison, Sally: Tulsa, December 17, 1986.
Duncan, Herbert: telephone interview, June 30, 1986.
Edmondson, Congressman Ed, Jr.: Muskogee, May 6, September 18, 1986.
Fields, Edythe Rambo: Muskogee, January 11, 1987.
Foss, R. E.: telephone interview, June 29, 1986.
Friman, Bill: Tulsa, March 3, May 13, 1986.
Gbeho, Ambassador Victor: New York, August 14, 1986.
Goodwin, Ed, Jr.: Tulsa, May 13, 27, September 15, 1986.
Goodwin, Mrs. E. L.: Tulsa, May 13, 1986.
Goodwin, Jim: Tulsa, May 13, 27, 1986.
Gootenberg, Roy: Washington, D.C., July 6, 1987.
Grayson, Minny: telephone interview, December 10, 1984.
Greer, Naomi: Muskogee, December 21, 1986.
Hannah, John: Muskogee, June 30, 1982; December 23, 1986.
Hall, Congressman Sam B., Jr.: telephone interview, March 23, 1987.
Harmon, Bill: telephone interview, May 1, 1987.

Harrison, Antoinette Fuhr: Muskogee, December 22, 1986.
Henry, Irene: telephone interview, December 3, 1986.
Hillhouse, Gordon: telephone interview, June 30, 1986.
Hines, Jewel: telephone interview, May 30, 1986.
Howell, Joseph E.: Tulsa, May 13, 1986.
Hudson, Lola: telephone interview, May 4, 1987.
Jamierson, Blanche: Battle Creek, Michigan, December 27, 28, 1985.
Jefferson, Dorothy: telephone interview, December 10, 1984.
Jefferson, Helen: Boynton, December 9, 1984.
Jeffrey, Pauline: Muskogee, January 11, 1987.
Jennings, Emery: Muskogee, May 29, 1986.
Johnson, Henry Aliva: Houston, Texas, May 15, 1986.
Jones, Congressman James R.: Washington, D.C., May 22, 1987.
Jordan, Alice: Tulsa, December 15, 1984.
Kerr, Mary B.: telephone interview, April 15, 1987
Keville, Maurice: Washington, D.C., July 6, 1987.
Lane, Jim: Muskogee, December 27, 1986.
Leake, James: Muskogee, December 23, 1986.
Lee, Dr. Robert E.: Accra, Ghana, July 19, 1986.
Littlefield, Daniel J., Jr.: telephone interview, December 18, 1984.
Livingston, Julius: Tulsa, May 13, 1986.
Lotsu, Austin: Accra, Ghana, July 18, 1986.
Luper, Clara: Oklahoma City, March 11, 1986.
Maitama-Sule, Ambassador Yusuff, telephone interview, August 10, 1982.
Massey, Alma: telephone interview, December 12, 1984.
McFarland, Donna, telephone interview, June 9, 1986.
McInturff, Florence Hart: telephone interview, May 28, 1986.
Montgomery, Doris: Poteau, Oklahoma, February 25, March 10, 1986.
Montgomery, Dr. John: Poteau, February 25, March 10, 1986.
Nelson, Milton: telephone interview, November 2, 1986.
Nigh, Governor George: Oklahoma City, March 11, 1986.
Olympio, Benito: London, August 5, 1986.
Omoyad, Issham: London, August 6, 1986.
Phillips, B. F. Jr.: telephone interview, September 29, 1986.
Pollack, Bill: Accra, Ghana, July 23, 1986.
Porter, State Senator Edward Melvin: Oklahoma City, June 2, 1986.
Rickards, Len M.: Bartlesville, Oklahoma, June 23, 1986.
Rummerfield, B. F.: Tulsa, December 22, 1986.
Ryan, Mrs. E. C.: Muskogee, December 25, 1986.
St. John, Bill: telephone interview, August 5, 1987.

Sanders, Mrs. M. L.: Holdenville, Oklahoma, March 8, June 1, 1986.
Shadeed, Ahmed: Tulsa, September 17, 1986; August 19, 1987.
Simmons, Annamarie: Muskogee, March 13, 1986.
Simmons, Barbara, Muskogee, June 3, 1983.
Simmons, Donald M.: Muskogee, June 29, 30, 1982; June 3, September 12, 14, 16, 24, 26, 27, October 17, 19, 1983; February 17, 24, March 13, May 9, 31, June 20, December 26, 1986; July 18, 1987; November 5, December 31, 1988.
Simmons, Donna R.: Muskogee, March 16, 1986.
Simmons, Eunice: Muskogee, September 21, 1983.
Simmons, Jake, III: Washington, D.C., December 13, 1985; March 31, May 30, August 14, September 24, 1986; January 4, 1989.
Simmons, Jim: Muskogee, September 21, 29, 1983.
Simmons, John W.: Oklahoma City, May 26, 1983; December 31, 1984; February 27, December 19, 1986; March 11, 1987; March 8, 1989.
Simmons, Johnnie Mae: Tulsa, Oklahoma, September 17, 1986.
Snacke, Chuck: Tulsa, June 19, 1986.
Sneden, Loring B.: telephone interview, September 6, 1986.
Stewart, James: Oklahoma City, May 9, June 3, 1986.
Swearingen, Wayne: Tulsa, March 12, 1986.
Tamakloe, Edith: Accra, Ghana, July 23, 1986.
Thomas, Owen: September 28, 1986.
Thompson, Willis H.: Tulsa, September 3, 1986.
Threat, Erma: Oklahoma City, February 27, March 8, December 20, 1986.
Van den Bark, Edwin: Bartlesville, Oklahoma, August 2, 1982; March 14, 1986.
Watts, The Reverend Wade: McAlester, Oklahoma, March 3, 10, 1986.
Webb, C. W.: Muskogee, Oklahoma, September 13, 1983.
Welch, Taft: Tulsa, March 15, 1982; February 19, 21, March 12, 1986.
Wesley, George: telephone interview, December 17, 1986.
Wheatley, Melba: Battle Creek, Michigan, December 27, 1985.
Wheeler, Clyde: telephone interview, May 28, 1986.
Zaroor, Fred: Muskogee, December 22, 1986.
Zarrow, Henry: Tulsa, November 18, 1986.
Zarrow, Jack: Tulsa, November 18, 1986.

Notes

Introduction

1. Doris Montgomery, interview.
2. Sylvan Amegashie, interview.
3. *Daily Graphic,* November 25, 1978.
4. Program notes for Annual Conference of Oklahoma State Conference of Branches NAACP, November 15, 1963, Box 129, Papers of the NAACP, Library of Congress.
5. J. J. Simmons III, interview.
6. Congressman James R. Jones, interview.
7. Interviews with Donald Simmons and J. J. Simmons III.

Prologue: AN ANCESTRY OF FREEDOM

1. Senate Committee on Indian Affairs, *Laws and Treaties,* 1904, 2: 932.
2. John Simmons, interview. Odie B. Faulk, *Muskogee: City and County* (Muskogee, Oklahoma: The Five Civilized Tribes Museum, 1982), 80; *Muskogee Daily Phoenix,* November 14 and 15, 1907.
3. Kaye M. Teall, ed. *Black History in Oklahoma: A Resource Book* (Oklahoma City: Oklahoma City Public Schools, 1971), 19–20; Daniel F. Littlefield, *Africans and Creeks: From the Colonial Period to the Civil War* (Westport, Connecticut, and London: Greenwood Press, 1979), 38–49; Angie Debo, *The Road to Disappearance: A History of The Creek Indians* (Norman, Oklahoma: University of Oklahoma Press, 1941), 44, 56;

W. E. B. Du Bois, *Black Reconstruction in America* (New York: Harcourt, Brace, 1935), 13.

4. Debo, *History,* 115–16.
5. Susie Martin Ross interview, Indian-Pioneer Papers, Vol. 61, 16, Western History Collections, University of Oklahoma, Norman; Littlefield, *Africans,* 46, 85, 145; Debo, *History,* 115.
6. Ned Thompson Interview, Indian-Pioneer Papers, Vol. 90, 387–88; Debo, *History,* 69, 115; Kenneth Porter Wiggins, "The Founder of the 'Seminole Nation' Secoffee or Cowkeeper," *The Florida Historical Quarterly,* April 1949, 362–84; John K. Mahon, *History of the Second Seminole War* (Gainesville, Florida: University of Florida Press, 1967), 16.
7. Debo, *History,* 26–31, 83.
8. Debo, *History,* 76–83; Mahon, *Seminole War,* 6.
9. Grant Foreman, *Indian Removal* (Norman, Oklahoma: University of Oklahoma Press, 1932), 107–18, 129–51; Debo, *History,* 100–101.

The pillage was so extreme that even federal officials protested. One government observer at the time noted: "I have never seen corruption carried on to such proportions . . . acts which should make men cover their face and shun daylight came to be the boast of these despoilers of Indian property . . . they seemed to carry on the business in sport; that a toast was given in a crowd, 'Here's to the man that can steal the most land tomorrow without being caught at it.' "

Nonetheless, the federal government used such fraud to coax the tribe to emigrate. In a little-known letter to Creek leaders in 1830, Jackson's Secretary of War John H. Eaton wrote:

> . . . the laws of the state (of Alabama and Georgia) are now extended over you, and your great father has not the power to prevent it. Daily are his red and white children coming nearer and nearer together until presently wars and strife and bloodshed will be the consequence. . . . The President directs me to say that in persuading you to remove West of the Mississippi, he feels that he is consulting your interest and happiness—nay more, your safety. . . . The soil [in the west] will be yours and the property of your children forever—no white man will be there to claim your grounds. . . . It is idle to console yourself with the hope that Congress will interfere in your behalf, those who tell you so intend to deceive you, Congress has no power over the subject. . . . When danger comes around you, will you appeal to your great father to afford protection? He will answer

> that he cannot because he has not the power. Will you unbury your tomahawk and go to battle? You are too weak and cannot do it.

Foreman, *Removal,* 119; Letter from Chief Roley McIntosh to President of the United States, August 28, 1841, containing copy of letter to Creek leaders from Secretary of War John Eaton, March 20, 1830. Records of the Office of Indian Affairs, National Archives. Microfilm Series M234, Roll 226, 290.

10. Debo, *History,* 88–90; Foreman, *Removal,* 21.
11. Foreman, *Removal,* 161, 179–80; Debo, *History,* 102; *American State Papers: Military Affairs,* 7:522; Q. L. Simmons's biography of Cow Tom, April 26, 1975, Historical Oklahoma Biographies Collection, Western History Collections, University of Oklahoma, Norman.
12. Kenneth Wiggins Porter: *The Negro on the American Frontier* (New York: Arno Press and *New York Times,* 1971), 238–61; "Negroes and the Seminole War, 1835–1842," *The Journal of Southern History,* November 1964, 427–50; "Florida Slave and Free Negroes in the Seminole War," *Journal of Negro History,* April 1943, 390–421. See also *Military Affairs,* 7: 821–835.
13. Mahon, *Seminole War,* 182; John W. Phelps, "Letters of Lieutenant John W. Phelps, U.S.A. 1837–1838," *Florida Historical Quarterly,* October 1927, 69; U.S. Department of the Interior, *Report of the Board of Indian Commissioners, 1870,* Appendix 37; *Military Affairs,* 7: 876.
14. Mahon, *Seminole War,* 298; *Indian Commissioners,* Appendix 37; Kenneth Wiggins Porter, "Negro Guides and Interpreters in the Early Stages of the Seminole War, December 28, 1835–March 6, 1837," *Journal of Negro History* (April 1950), 177; Major Thomas Childs, "Extracts from His Correspondence with his Family," *Historical Magazine,* 3rd. ser., Vol. 3, 170.
15. Mahon, *Seminole War,* 78.
16. Porter, "Negro Guides and Interpreters," 177; Childs, Extracts, 170; Mahon, *Seminole War,* 204; *Military Affairs,* 7: 838–41; Samuel Forry, "Letters of Samuel Forry, Surgeon, U.S. Army, 1837–1838," *Florida Historical Quarterly,* January 1928, 134–35.
17. Foreman, *Removal,* 184–90; Susie Martin Ross interview, Indian-Pioneer Papers, 16; U.S. Bureau of the Census, *Indian Lands West of Arkansas (Oklahoma), Population Schedule of the United States Census of 1860,* transcribed by Frances Woods (Oklahoma City: Oklahoma Historical Society, 1964), 6–7.
18. Debo, *History,* 103, 108–12.

19. Debo, *History,* 106; Littlefield, *Africans,* 116, 154; Sigmund Sameth, "Creek Negroes: A Study of Race Relations" (M.A. thesis, University of Oklahoma, 1940), 92; Letter from Seminole Agent J. W. Wright to General Howard, February 10, 1867, National Archives, "Letters from Creek Agency," M234, Roll 231, document W44.
20. Susie Martin Ross interview, Indian-Pioneer Papers, 16; Littlefield, *Africans,* 139, 150.
21. Susie Martin Ross interview, Indian-Pioneer Papers, 16; Littlefield, *Africans,* 138.
22. Littlefield, *Africans,* 138–39; Debo, *History,* 115; Siegal E. McIntosh interview, Indian-Pioneer Papers, vol. 35, 236–37; Senate Committee on Indian Affairs, *Claims of the Loyal Creeks,* 57th Cong., 1st sess., 1902, S. Rept. 420, 35; Sameth, *Creek Negroes,* 23.
23. Grant Foreman, ed., *A Traveler in Indian Territory* (Cedar Rapids, Iowa: The Torch Press, 1930), 148.
24. Littlefield, *Africans,* 139; David Barnet's blacksmith contract with Creek agent, February 19, 1840; letters from Creek agency, M234–226: 336.
25. Claim 160, *Records Relating to Loyal Creek Claims, 1869–70,* Records of the Bureau of Indian Affairs, National Archives.
26. J. W. Wright to General Howard, M234–231: W44.
27. Debo, *History,* 110, 114, 125–27, 142–60; Foreman, *Traveler,* 112; Littlefield, *Africans,* 138, 233–38; John Bartlett Meserve, "Chief Opotheayahola," *Chronicles of Oklahoma,* Winter 1931–32, 439–53; Teall, *Black History,* 54–56.
28. Claim 160, *Loyal Creek Claims;* Teall, *Black History,* 63–64; Debo, *History,* 155; Lary C. Rampp, "Negro Troop Activity in Indian Territory, 1863–1865," *Chronicles of Oklahoma,* Spring 1969, 541–48.

    Confederate troops expected the blacks to break under fire and surrender (their supplies included cases of shackles to chain the former slaves and return them to captivity). They were unpleasantly surprised. The five hundred men from the Kansas First Colored Volunteers, many of them refugees from Indian Territory, tipped the battle scales. The Union's commanding general, who had used the blacks for his deadly front-line charge, reported, "Their coolness and bravery I have never seen surpassed."
29. Claim 160, *Loyal Creek Claims;* Teall, *Black History,* 72.
30. U.S. Department of the Interior, *Report of the Commissioner of Indian Affairs, 1863,* 223–24; *Report of the Commissioner, 1864,* 342–44; Littlefield, *Africans,* 240; Anne Heloise Abel, *The American Indian Under Reconstruction* (Cleveland, Ohio: Arthur H. Clark, 1925), 272.
31. Mahon, *Seminole War,* 91; U.S. Senate, *Loyal Creeks,* 20.

32. John Bert Campbell, *Campbell's Abstract of Creek Freedman Census Cards and Index* (Muskogee, Oklahoma: Oklahoma Printing Co., 1915), Card No. 127.
33. Littlefield, *Africans,* 239; Debo, *History,* 161.
34. *Report to the Commissioner, 1864,* 342–44; Littlefield, *Africans,* 240–41; Claim 160, *Loyal Creek Claims;* Debo, *History,* 162–63.
35. U.S. Department of War, *War of Rebellion: Official Records of the Union and Confederate Armies,* 1st ser., 41: 456f., 542f., 870–73; Debo, *History,* 165, 177; M. Thomas Bailey, *Reconstruction in Indian Territory* (Port Washington, New York: Kennikat Press, 1972), 40–41.
36. *Report of the Commissioner, 1865,* 314, 318–19; Debo, *History,* 165–69, 175; Littlefield, *Africans,* 241–42; Gail Balman, "The Creek Treaty of 1866," *Chronicles of Oklahoma,* Summer 1970, 184–88; Bailey, *Reconstruction,* 66–67, 76–77.
37. *Report of the Commissioner, 1865,* 312–41; Balman, *Chronicles,* 192–93; Debo, *History,* 169.
38. President Lincoln's initial solution to the "Negro Problem" was an attempt to persuade four million free blacks to leave the country and form black colonies overseas. In early 1864, he convinced the government of Haiti, one of the world's first free black republics, to allow northern promoters to try a resettlement attempt. They moved hundreds of freedmen to an island off the Haitian mainland, but the experiment backfired when the freedmen were horribly exploited. Half of them died before the U.S. government could bring them home.

    Richard Hofstadter, ed. *The United States 1916–1970: A World Power,* 4th ed. (Englewood Cliffs, New Jersey: Prentice-Hall, 1976), 315–16; John Hope Franklin, *From Slavery to Freedom: A History of Negro Americans* (New York: Alfred A. Knopf, 1947), 240.
39. Hofstadter, *World Power,* 311, 318–19; Franklin, *Freedom,* 241–43.
40. *Report of the Commissioner, 1865,* 341.
41. The commissioner of Indian affairs claimed that by enfranchising their former slaves, the tribes would "be following the example of the white people of the United States who have, from the beginning, admitted to the rights of citizenship white people of all countries in the world, where there has appeared to exist no natural antagonism; that as a result of this policy the whites have grown so numerous and strong as to render it difficult for the President to prevent them crushing out the Indian race and that many of the States, including the richest and wisest, make no distinction in this respect on account of color."

    Abel, *American Indian,* 275–76.
42. Abel, *American Indian,* 273–75; *Report of the Commissioner, 1866,* 283–84.

43. Debo, *History*, 171–73, 177–78; Bailey, *Reconstruction*, 75–76; Letter from D. N. McIntosh to Commissioner Dennis Cooley, March 18, 1866, M234–231: M245.
44. Debo, *History*, 171–75; Turner, C. W., "Events Among the Muskogees During Sixty Years," *Chronicles of Oklahoma*, March 1932, 26.
45. *Report of the Commissioner, 1866*, 10.
46. *Laws and Treaties*, 932–33.
47. Sameth, *Creek Negroes*, 31.
48. Some Oklahomans viewed the enfranchisement as a scam. In 1930 Clarence W. Turner, a prosperous white settler from territorial days, wrote, "The northern Creeks had as their interpreter an old freed slave named Harry Island. Harry was sharp as tacks and made most of his position. He got the government representative to put a clause in the treaty that resulted in the Creeks having to give each colored individual and descendants equal shares of their land and money. A greater piece of robbery never was imposed on a helpless people and some day the government should reimburse these Creek Indians."

    Turner, *Chronicles*, 26.
49. *Report of the Commissioner, 1868*, 285; Bailey, *Reconstruction*, 69–72, 79; Gary R. Kremer, "For Justice and a Fee: James Milton Turner and the Cherokee Freedmen," *Chronicles of Oklahoma*, Winter 1980–81, 377–91.
50. *Report of the Commissioner, 1866*, 318–19.
51. *Report of the Commissioner, 1870*, Appendix 37.
52. *Report of the Commissioner, 1870*, Appendix 37; John Harrison interview, Indian-Pioneer Papers, Vol. 5, 324–52.
53. Letter from Superintendent W. Wyers to Commissioner Bogy, January 1, 1867, M234–236: B34½.
54. Letter from Acting Commissioner C. E. Mix to Secretary of the Interior O. H. Browning, November 16, 1867, Report Books Vol. 17, Report of the Bureau of Indian Affairs, Record Group 75, National Archives; letter from Creek Agent J. W. Dunn to Commissioner of Indian Affairs, June 1, 1868, M234–236: D1195.
55. A new democratic system was initiated after the Civil War which allowed each tribesman to vote for members of Congress and one principal chief. Checote's followers set up the Electoral College–like system, but never taught the Upper Creeks how to use it. When the tribe held its first election in November 1867, the black and loyalist majority supporting Chief Sands attempted to vote as they always had: appearing in the public square, lining up behind their candidate, and waiting to be counted. They never thought to cast a ballot. Checote's supporters did, and he won in a landslide. Chief Sands and his supporters felt they had been cheated out

of power. They protested the election and ignored the new constitution, creating a postwar rift that continued into the next century.

Debo, *History*, 179–91.

56. Letter of the Secretary of the Interior, 40th Con., 21 sess., 1868, S. Ex. Doc. 64; letter from J. W. Wright to General Howard, M234–231: W44; letter from James Marthaus, Superintendent of Indian Affairs to N. G. Taylor, Commissioner of Indian Affairs, August 31, 1867, M234–231: W 427.
57. Hofstadter, *World Power*, 321–22; Indian-Pioneer Papers, Vol. 5, 353, Western History Collections, University of Oklahoma, Norman.
58. *Congressional Globe*, 40th Cong., 2d sess., 1868, 39, Part 3: 2595.
59. Debo, *History*, 178; Bailey, *Reconstruction*, 104; letter of Secretary of the Interior, 40th Cong., 2d sess., 1868, S. Ex. Doc. 64.
60. Frederick Douglass, West India Emancipation Speech of August 1857, cited in Arnold Adoff, ed., *Black on Black* (New York: Macmillan, 1968), 1.
61. *Oklahoma Eagle*, July 26, 1984.

Chapter 1 CHILD OF THE FRONTIER

1. Interviews with Ophelia Brown, Donald Simmons, and Erma Threat.
2. Interviews with Ophelia Brown, Johnnie Mae Simmons, and Erma Threat.
3. Interviews with John Simmons and Johnnie Mae Simmons; John Bert Campbell, *Campbell's Abstract of Creek Freedman Census Cards and Index* (Oklahoma Printing: Muskogee, Oklahoma, 1915); Jim Tomm interview, Pioneer History, Vol. 112, 294; Debo, *History*, 17; Pay Roll of Creek National Council, October 1877, Creek Tribal Records #32563, Oklahoma Historical Society, Oklahoma City; letter certifying John Jefferson's election to House of Warriors, August 16, 1897, Creek Tribal Records #33439.
4. John Simmons, interviews; biography of Jake Simmons, Sr., dictated to Pearl Phillips Simmons, unpublished and undated family document.
5. Interviews with Bill Friman, Florence Hart McInturff, Donald Simmons, John Simmons, and Erma Threat; Campbell, *Abstract;* biography of Jake Simmons, Sr., by Jake Simmons, Jr., unpublished and undated family document; Jake Simmons, Sr., interview, Pioneer History, Vol. 9, 344–45.

6. Jake Simmons, Sr., interview, Pioneer History, Vol. 9, 354; biography of Jake Simmons, Sr., by Jake Simmons, Jr.; biography of Jake Simmons, Sr., dictated to Pearl Phillips Simmons; interviews with Ophelia Brown, Florence McInturff, and John Simmons; Okmulgee Historical Society, *History of Okmulgee County, Oklahoma* (Tulsa: Historical Enterprises, 1985), 980–84.
7. Debo, *History,* 261; Jake Simmons, Sr., interview, Pioneer History, Vol. 9, 344–410.
8. Teall, *Black History,* 108–26; Debo, *History,* 253–57; William Loren Katz, *The Black West* (Garden City, New York: Doubleday, 1971), 139–42, 200–12; D.C. Gideon, *Indian Territory* (New York: Lewis Publishing, 1901), 115; Nudie E. Williams, "Black Men Who Wore the Star," *Chronicles of Oklahoma,* Autumn 1983, 83–86; Carolyn Thomas Foreman, "The Light-Horse in the Indian Territory," *Chronicles of Oklahoma,* Spring 1956; William H. Leckie, *The Buffalo Soldiers* (Norman, Oklahoma: University of Oklahoma Press, 1979), 12–28.
9. John Simmons, interviews; Debo, *History,* 228; Jake Simmons, Sr., interview, Pioneer History, Vol. 9, 393–94.

   Those found guilty of serious crimes like theft were tied to a whipping post, lectured, and given twenty-five lashes with a hot hickory stick. Second offenders got fifty lashes. The third offense—and the crime of murder—were punishable by death. While firing-squad executions occurred infrequently they were major public events; the condemned got to choose his coffin and the type of gun used to kill him.
10. Interviews with Ophelia Brown, Donald Simmons, John Simmons, and Erma Threat.
11. Biography of Jake Simmons, Sr., dictated to Pearl Simmons, 14–18.
12. Interviews with Jim Simmons, John Simmons, and Erma Threat.
13. Interviews with Jim Simmons, John Simmons, and Erma Threat; Campbell, *Abstract,* 26, 37.
14. Interviews with Ophelia Brown, Sam Brown, John Simmons, Johnnie Mae Simmons, and Erma Threat; Jake Simmons, Sr., "From Cowboy to Landlord," *Southern Workman,* March 1915, 177.
15. Interviews with Ophelia Brown, Sam Brown, and John Simmons; Simmons, *Southern Workman,* 176; Booker T. Washington, *Up from Slavery* (New York: Doubleday, 1901), xxxi.
16. John Simmons, interviews.
17. John Simmons, interviews.
18. Interviews with Ophelia Brown, John Simmons, and Erma Threat; Simmons, *Southern Workman,* 177.

19. Interviews with Ophelia Brown, Donald Simmons, John Simmons, and Erma Threat.
20. Interviews with Ophelia Brown, Donald Simmons, John Simmons, Johnnie Mae Simmons, and Erma Threat.
21. Interviews with Ophelia Brown, John Simmons, and Erma Threat; interview with Jake Simmons, Sr., Pioneer History, 373.
22. Jake Senior was a big talker. In 1937 a WPA worker interviewed him for the government's extensive Indian Pioneer History project. More than a thousand Oklahomans were included in the 140-volume work. Only a handful talked as much as Simmons, whose transcript ran sixty-five pages. Afterward his interviewer commented, "Jake Simmons is a very interesting man to talk to. He is not only familiar with the happenings of the times a half century or more ago but is right up to the minute at the present time. There is nothing more he enjoys than to contact his boyhood friends and talk over the days when they were boys. . . . His education is limited but his practical knowledge of affairs has made him a great character."

    Interview with Jake Simmons, Sr., Pioneer History, 410.
23. Erma Threat, interviews.
24. Interviews with Johnnie Mae Simmons and Erma Threat.
25. Biography of Jake Simmons, Sr., dictated to Pearl Simmons, 19–20; John Simmons, interviews.
26. Interviews with Ophelia Brown, John Simmons, and Johnnie Mae Simmons; *Oklahoma Eagle,* December 8, 1955.
27. Interviews with Ophelia Brown, Donald Simmons, and John Simmons; Debo, *History,* 231; Angie Debo, *And Still the Waters Run: The Betrayal of the Five Civilized Tribes* (New York: Alfred A. Knopf, 1944), 133; Jimmie Lewis Franklin, *Journey Toward Hope* (Norman, Oklahoma: University of Oklahoma Press, 1982), 25.
28. Interviews with John Simmons, Johnnie Mae Simmons, and Erma Threat.
29. Debo, *History,* 221, 325, 332f., 362f., 370–73; Debo, *Waters,* 13, 55, 152.
30. Nonetheless, celebrated Creek historian Angie Debo blamed the blacks for breaking up the Indian nation. In her 1941 book, *The Road to Disappearance,* she wrote: "The Creeks, of course, had invited this result by the laxity of their own policy. Characteristically they had offered an asylum to a deeply wronged people, and only a few of their Southern leaders had foreseen the oblique but constant aggressiveness of this humble and pliant race. The full bloods discovered their mistake in the 1890s, when it was too late to correct it. The Negro influence within their

tribe had been second only to the encroachment of the white man in the destruction of their nationality."

Debo, *History*, 371–72.

31. Debo, *History*, 347, 372.
32. Debo, *Waters*, 45–51; Debo, *History*, 370.
33. Franklin, John Hope, *From Slavery to Freedom: A History of Negro Americans* (New York: Alfred A. Knopf, 1947), 236, 248; Du Bois, *Reconstruction*, 368, 601, 603; Debo, *Waters*, 45–51; U.S. Bureau of the Census, Part 1 of *Census of the United States: 1900, Agriculture*, Vol. 5 (Washington, D.C.: U.S. Government Printing Office, 1902); U.S. Bureau of the Census, *Census of the United States, 1910, Agriculture*, Vol. 7 (Washington, D.C.: U.S. Government Printing Office, 1913).
34. Du Bois, *Reconstruction*, 602, 611.
35. Debo, *Waters*, 45–51.
36. Debo, *Waters*, 92, 126–37; Sigmund Sameth, "Creek Negroes: A Study of Race Relations" (M.A. thesis, University of Oklahoma, 1940), 46, 54, 92; *Magnum Star*, May 5, 1904; *Oklahoma City Times*, November 20, 1908.
37. Debo, *Waters*, 144.
38. Interviews with Sam Brown and Erma Threat.
39. Simmons ran into the shrewd financier after a number of years, at a bank reception for a Creek rancher he once worked for. The rancher asked how he had been getting along. The banker, standing nearby, answered for him: "Jake's doing fine—he got by." It suddenly dawned on Simmons that the banker, whom he had always assumed to be a trustworthy ally, had all along been waiting to take over his ranch. The experience left a deep impression on Simmons. From that day on he periodically advised children and grandchildren, "Son, you got to get by."

    Donald Simmons, interview.
40. Hastains Index for Creek Nation, Gilcrease Museum, Tulsa, Oklahoma.
41. Simmons, *Southern Workman*, 176–77; interviews with Donald Simmons, John Simmons, and Erma Threat.

Chapter 2 "I WANT TO BE AN OILMAN"

1. Washington, *Up from Slavery*, 281.
2. Franklin, *Freedom*, 252–69; William Loren Katz, *Black People Who Made the Old West*, 152; Carter G. Woodson, *A Century of Negro Migration* (New York: Russell and Russell, 1969), 134–46; William Windom, "The

Senate Report on the Exodus of 1879," *Journal of Negro History*, January 1919; Roy Garvin, "Benjamin or 'Pap' Singleton and His Followers," *Journal of Negro History*, January 1948.

3. Teall, *Black History*, 150–66; Franklin, *Journey*, 10–16; Arthur Lincoln Tolson, *The Black Oklahomans, A History: 1541–1972* (New Orleans: Edwards Printing, 1974), 41–46, 75–85, 94; Norman L. Crockett, *The Black Towns* (Norman, Kansas: Regents Press of Kansas, 1979), 16–26; Jane W. Roberson, "Edward P. McCabe and the Langston Experiment," *Chronicles of Oklahoma*, Spring 1969; Berlin B. Chapman, "Freedmen and the Oklahoma Lands," *The Southern Social Science Quarterly*, September 1948; Nudie E. Williams, "The Black Press in Oklahoma: The Formative Years, 1889–1907," *Chronicles of Oklahoma*, Autumn, 1983, 310.
4. *New York Times*, March 1, 1890.
5. *Kansas City Evening News*, March 4, 1890; U.S. Department of Commerce, Part 1 of *Historical Statistics of the United States: Colonial Times to 1970* (Washington, D.C.: U.S. Government Printing Office, 1975), 33.
6. Teall, *Black History*, 167, 172–79; 202–204; Crockett, *Black Towns*, 98; Franklin, *Journey*, 33; Paul Mellinger, "Discrimination and Statehood in Oklahoma," *Chronicles of Oklahoma*, Autumn 1971; *Muskogee Cimeter*, February 23, 1905.
7. Jame M. Smallwood, "The Black Experience in Oklahoma" (Ph.D. diss., Oklahoma State University, 1981), 9: Edwin C. McReynolds, *Oklahoma: A History of the Sooner State* (Norman, Oklahoma: University of Oklahoma Press, 1954), 316; Mellinger, *Chronicles*, 359, 366–73.
8. Sameth, *Race Relations*, 61–62, 70; Kenneth Wiggins Porter, *The Negro on the American Frontier* (New York: Arno Press and the *New York Times*, 1971), 79.
9. John Simmons, interviews; Mellinger, *Chronicles*, 373–74; Crockett, *Black Towns*, 93–98; Teall, *Black History*, 239.
10. Teall, *Black History*, 167–68; Crockett, *Black Towns*, 35, 84, 146–49, 181.
11. Sameth, *Race Relations*, 56.
12. Sameth, *Race Relations*, 50.
13. Interviews with John Simmons and Erma Threat.
14. John Simmons, interviews.
15. Crockett, *Black Towns*, 168–69.
16. Teall, *Black History*, 286–88.
17. Salim Muwakkil, "Marcus Garvey: Black Moses Celebrated by Aged Followers," *In These Times*, September 10–16, 1986; Marcus Garvey, *Philosophy and Opinions of Marcus Garvey*, Vols. 1 and 2 (New York: Atheneum, 1986); Donald Simmons, interviews.
18. Interviews with Donald Simmons and J. J. Simmons III.

19. Carl Coke Rister, *Oil Titan of the Southwest* (Norman, Oklahoma: University of Oklahoma Press, 1949), 12–13; Dr. Joseph Faust and John F. Hassler, commissioned appraisal of oil in Creek lands, April 3, 1975, 11–14, 20, Special Collections, Northeastern State University, Tahlequah, Oklahoma; Robert Gregory, *Oil in Oklahoma* (Muskogee, Oklahoma: Leake Industries, 1976); Scott Ellsworth, *Death in a Promised Land: The Tulsa Race Riot of 1921* (Baton Rouge, Louisiana: Louisiana State University Press, 1982), 9–10.
20. Gregory, *Oil,* 14; Ellsworth, *Death,* 9; Rister, *Titan,* 15; Debo, *Waters,* 87–88, 133.
21. Interviews with Donald Simmons, J. J. Simmons III, John Simmons, and Johnnie Mae Simmons; Simmons, *Southern Workman,* 177.
22. Interviews with Ophelia Brown, Johnnie Mae Simmons, and Erma Threat.
23. Erma Threat, interviews.
24. John Simmons, interviews.
25. Interviews with Espanola Barnes and Erma Threat.
26. Interviews with John Simmons and Erma Threat. Teall, *Black History,* 89; Henry S. Myers interview, Indian and Pioneer History, Vol. 65, 579, Western History Collections, University of Oklahoma, Norman.
27. Address of Booker T. Washington at Muskogee Convention Hall, August 19, 1914, Booker T. Washington Papers, Library of Congress.
28. Hugh Hawkins, *Booker T. Washington and His Critics* (Lexington, Massachusetts: D. C. Heath & Co., 1974), 49, 61, 187–92.
29. Washington's roster of "sainted philanthropists" read like a *Who's Who* of northern capitalists. Men like George Eastman, Andrew Carnegie, John D. Rockefeller, Jacob Henry Schiff, and Julius Rosenwald contributed hundreds of thousands of dollars to Tuskegee. Because he was seen as a man who could keep his people in line Washington was afforded immense power by the system. Using a network of supporters, agents, and informers, "The Wizard of Tuskegee" made sure that virtually all the appointments made under Presidents Theodore Roosevelt and William Howard Taft went to devoted supporters or sympathizers. Like most people with power Washington was no saint. He had a vengeful streak; one was either with him or against him. He blackballed opponents and employed spies to harass them. In his day Washington achieved a position of dominance in the black community which no person has ever equaled.

    Hawkins, *Critics,* 48, 185; Washington, *Up from Slavery,* xxxix; Lerone Bennett, Jr., *Before the Mayflower* (Chicago: Johnson Publishing, 1975), 274–77; Louis B. Harlan, *Booker T. Washington: The Wizard of*

Tuskegee, 1901–1915 (New York and Oxford: Oxford University Press, 1983), 17, 130, 376–78.

30. *Muskogee Daily Phoenix,* November 17, 1907, August 18 and August 19, 1914; *Muskogee Times-Democrat,* August 19 and 20, 1914; *Tulsa Star,* August 29, 1914; Harlan, *Wizard,* 417.
31. Address of Booker T. Washington, August 19, 1914, Papers.
32. *Muskogee Times-Democrat,* August 21, 1914; *Muskogee Daily Phoenix,* August 21, 1914; *Tulsa Star,* August 29, 1914; interviews with Jim Simmons and John Simmons.
33. Interviews with John Simmons and Erma Threat.
34. Washington, *Up from Slavery,* 281; John Simmons, interview.
35. John Simmons, interviews. Jake Simmons, Jr.'s, academic records, Office of the Registrar, Tuskegee University, Tuskegee, Alabama.

Chapter 3 LEARNING FROM THE WIZARD

1. Booker T. Washington, *Up from Slavery* (Garden City, New York: Doubleday, 1901), 55.
2. Louis R. Harlan, *Booker T. Washington: The Wizard of Tuskegee 1901–1915* (New York and Oxford: Oxford University Press, 1983), 144–45; Washington, *Up from Slavery,* 45–59; Tuskegee Normal and Industrial Institute, *School Catalog 1914* (Tuskegee, Alabama: Tuskegee Institute, 1914), 17–21; Milton Nelson, interview.
3. Harlan, *Wizard,* 148.
4. Reverend Anson Phelps Stokes, *Tuskegee Institute: The First Fifty Years* (Tuskegee, Alabama: Tuskegee Institute Press, 1931), 91; Tuskegee Institute, *Tuskegee Institute Centennial Celebration* (Tuskegee, Alabama: Tuskegee Institute Press, 1981), 7–13; Harlan, *Wizard,* 149–55, 164; Washington, *Up from Slavery,* 45–59, 148–52; Tuskegee, *Catalog,* 17–21; interviews with Milton Nelson and John Simmons.
5. Harlan, *Wizard,* 155, 163.
6. Harlan, *Wizard,* 144 –57; Tuskegee, *Catalog,* 17–21; interviews with Milton Nelson, John Simmons, and Melba Wheatley.
7. Tuskegee, *Catalog,* 17–21; Washington, *Up from Slavery,* 174–76.
8. Tuskegee University, *Academic Records 1914–1919;* Tuskegee, *Catalog,* 27.
9. Donald Simmons, interviews.

   "There is no education which one can get from books and costly apparatus," Washington said, "that is equal to that which can be gotten from contact with great men and women." Washington, *Up from Slavery,* 55.

10. Donald Simmons, interviews; John Bartlett, *Familiar Quotations* 15th ed. (Boston: Little, Brown, 1980), 686.
11. Interviews with Donald Simmons, J. J. Simmons III, John Simmons, and Melba Wheatley.
12. Interviews with Donald Simmons, John Simmons; Harlan, *Wizard,* 144.
13. Interviews with Milton Nelson, Donald Simmons, John Simmons, Melba Wheatley; Tuskegee Institute, Classes of 1909 and 1919, *Reunion Program,* 1959.
14. Interviews with Donald Simmons and John Simmons.
15. Washington, *Up from Slavery,* 218–22.
16. Harlan, *Wizard,* 359–78.
17. Harlan, *Wizard,* 164–72; Stokes, *Fifty Years,* 30.
18. Harlan, *Wizard,* 172 (original citation in *Mobile Herald,* October 13, 1903).
19. Harlan, *Wizard,* 171.
20. Lerone Bennett, Jr., "Chronicles of Black Courage, Part V." *Ebony* (September 1983), 131–34; Tuskegee, *Centennial,* 8.
21. Donald Simmons, interviews.
22. Donald Simmons, interviews.
23. Harlan, *Wizard,* 438–55.

    Washington was in New York when he collapsed from nervous exhaustion and arteriosclerosis. After doctors told him he had only a few days to live he dragged himself out of a hospital bed and hurried to Penn Station, where he refused a wheelchair, leaning instead on his wife, Margaret, and caught the next train to Alabama. His doctor noted that it was "uncanny to see a man up and about who ought by all the laws of nature to be dead." Washington made his last public statement before leaving New York. "I was born in the South, I have lived and labored in the South, and I expect to die and be buried in the South," he said. He lost consciousness when he got to Tuskegee and died at home the next morning.
24. Washington, *Up from Slavery,* 244; interviews with Donald Simmons and Donna Simmons; Jake Simmons, Jr.'s, speech, "The Economic Status of Negroes in the United States," as broadcast by the Voice of America, December 4, 1963; Jake Simmons, Jr.'s, recorded conversation with Naomi Greer, late 1980.
25. Donald Simmons, interviews.

26. Robert Lacey, *Ford: The Men and the Machine* (Boston and Toronto: Little, Brown, 1986), 169–72, 218–23; Donald Simmons, interviews.
27. Donald Simmons, interviews.
28. Harlan, *Wizard,* 130; interviews with Donald Simmons and John Simmons.
29. Interviews with Donald Simmons and John Simmons.
30. Interviews with Donald Simmons and John Simmons.
31. Stokes, *Fifty Years,* 31; Tuskegee, *Celebration,* 5; Washington, *Up from Slavery,* 267–68; Harlan, *Wizard,* 152, 270+.
32. Interviews with Blanche Jamierson, J. J. Simmons III, John Simmons, and Melba Wheatley.
33. Interviews with Blanche Jamierson, Donald Simmons, John Simmons, and Melba Wheatley; Jake Simmons, Jr.'s, speech, "The American Negro as a Soldier," May 22, 1918.
34. *Thirty Years of Lynching in the United States: 1889–1918* (New York: National Association for the Advancement of Colored People, 1919), 30; Jake Simmons, Jr.'s, speech, "The League of Nations," 1918.
35. Jake Simmons, Jr.'s, speech, untitled, ca. 1918.
36. Harlan, *Wizard; Tuskegee Institute Commencement Guide,* 1919.
37. *Tuskegee Institute Commencement Guide,* 1919; Jake Simmons, Jr.'s, speech, "Class Creed," May 21, 1919.
38. Interviews with John Simmons and Melba Wheatley.
39. Jake Simmons, Jr.'s, recorded conversation with Naomi Greer, late 1980; interviews with J. J. Simmons III and Melba Wheatley.
40. Interviews with Blanche Jamierson, J. J. Simmons III, Erma Threat, and Melba Wheatley.
41. Melba Wheatley, interview.

    The decision probably had something to do with his father's influence. Jake came from a southwestern tradition, where women supported their men and reared children in a stable home. Melba's mother wasn't like this, and neither was she. Indeed, the very instability which Jake Sr. warned his son against came to pass. Melba was married five times in her life, the second very soon after Jake returned to Muskogee. "I just decided that I'd best get married again," she recalls, "to show him I could get married, I guess." Although she led an exciting, full life, it was not that of a southern housewife and mother. Like her parents Melba spent many of her adult years on the road as a singer. Her children spent much of their adolescence with their grandmother.
42. Interviews with Blanche Jamierson, Donald Simmons, J. J. Simmons III, and Melba Wheatley.

43. Interviews with Blanche Jamierson, Donald Simmons, J. J. Simmons III; Cuyahuga County marriage certificate, December 20, 1920.
44. Washington, *Up from Slavery,* 219–20.

Chapter 4 THE LEASE HOUND

1. Taft Welch, interview.
2. The only relative who remained a farmer for life was Jake Sr.'s grandson Gleaton. Gleaton, who is Erma Threat's brother, still lives on a section of his family's land with their younger brother Sam. At the ages of seventy-seven and sixty-nine the pairs' hillbilly existence provides an unusual look into the world their cousins and uncles left behind. I visited them during the rainy season; their home resembled a tiny clapboard island in an ocean of muddy earth. The brothers manage a small herd of fifty beef cattle and farm a few patches of greens, turnips, potatoes, beans, and okra. Their modest life seems transfixed in time, and their limited exposure to the modern world bespeaks an age gone by. Gleaton and Sam's sparse living room has a few records lying out on a dusty bureau, almost as decorations (most conspicuous is a how-to album entitled "Put Down Your Whiskey Bottle"), but there seems to be no stereo to play them on. When Gleaton listened to a playback of his taped interview, his face grew excited and he yelled, "That's me!" like a man who had never heard his voice recorded before. "That's sure you talking!" Sam chimed in, eager to forward the tape to hear his own voice. The brothers are aware that they are anachronisms in 1989. Gleaton laments that young men today don't want to do ranch work—he's had to stop branding his cattle because "you can't hardly find no one to work for you." But neither of them thinks of leaving. "You know how far I traveled in my life?" Gleaton asked me, quite unexpectedly, as I left for cities far beyond. "I never passed Tulsa."
3. Donald Simmons, interviews; Oklahoma State Election Board, *Directory of Oklahoma, 1981* (Norman, Oklahoma: University of Oklahoma Printing Services, 1981), 555, 610, 617; "Okie from Muskogee," words and music by Merle Haggard and Roy Edward Burris. Copyright © 1969 Tree Publishing Co., Inc. All Rights Reserved. International Copyright Secured. Used by Permission of the Publisher.
4. Interviews with Cleora Butler, Donald Simmons, J. J. Simmons III, John Simmons, and Erma Threat.

5. Interviews with Donald Simmons and J. J. Simmons III; Jimmie Lewis Franklin, *Journey Toward Hope: A History of Blacks in Oklahoma* (Norman, Oklahoma: University of Oklahoma Press, 1982), 25.
6. Interviews with Donald Simmons, J. J. Simmons III, John Simmons, Willis H. Thompson, Erma Threat, and Taft Welch.
7. Interviews with Donald Simmons, J. J. Simmons III; Angie Debo, *And Still the Waters Run: The Betrayal of the Five Civilized Tribes* (New York: Alfred A. Knopf, 1944), 273–314; *Tulsa Star,* January 18, 1916; *Muskogee Daily Phoenix,* August 7–9, 1918.
8. Interviews with Donald Simmons and J. J. Simmons III.
9. Scott Ellsworth, *Death in a Promised Land: The Tulsa Race Riot of 1921* (Baton Rouge, Louisiana: State University Press, 1982), 9, 16, 22, 25–33, 99–100; Franklin, *Journey,* 142; *New York Times,* July 1, 1923; Taft Welch, interview.

   The mobs also went after leftists. In 1917 twelve members of the International Workers of the World (IWW) were sentenced to jail for organizing oilfield workers into a union. On the way to prison the men were conducted by fifty armed men in black robes, among them the chief of police. The organizers were savagely whipped, had hot tar and feathers poured on their bloodied backs, and were then run out of town. In an editorial headlined "Get Out the Hemp" the *Tulsa World* (still the city's largest paper) captured the violence of the period by applauding the mob as a "patriotic party" and recommending a bolder course of action: "If the IWW or its twin brother, the Oil Workers Union, gets busy in your neighborhood, kindly take occasion to decrease the supply of hemp. A knowledge of how to tie a knot that will stick might come in handy in a few days. . . . Kill 'em just as you would kill any kind of snake. Don't scotch 'em; kill 'em. And kill 'em dead. It is no time to waste money on trials and continuances like that. All is necessary is the evidence and a firing squad."
10. Ellsworth, *Death,* 128; Franklin, *Journey,* 135; interviews with Donald Simmons and Eunice Simmons.

    Eunice Simmons, Jake's sister-in-law, remembers a confrontation during her childhood which followed rumors that a mob was about to lynch a black man in the Muskogee jail. Her father, a minister, helped organize a band of Black Creeks who surrounded the area with high-powered rifles. "They didn't have any fear of white people," says Eunice Simmons. "That night there were a thousand black men in the north part of town, saying if this man goes out of jail tonight we will tear up Muskogee—as long as we can shoot we will kill that many people. And of course that man wasn't lynched."

11. Franklin, *Journey,* 136.
12. Ellsworth, *Death,* 24; *New York Times,* June 5, 1921.
13. Ellsworth, *Death,* 45–70, 103, 127; Franklin, *Journey,* 144–49; Kaye M. Teall, ed., *Black History in Oklahoma: A Resource Book* (Oklahoma City: Oklahoma City Public Schools, 1971), 204–208; Edwin C. McReynolds, *Oklahoma: A History of the Sooner State* (Norman, Oklahoma: University of Oklahoma Press, 1954), 338–39; *Tulsa Tribune,* June 2, 1971.
14. Sheldon Neuringer, "Governor Walton's War on the Ku Klux Klan: An Episode in Oklahoma History," *Chronicles of Oklahoma,* Summer 1967, 153–79; Ellsworth, *Death,* 102–103; Teall, *Black History,* 210; McReynolds, *Oklahoma,* 343–49; Langston Hughes, *Fight for Freedom: The Story of the NAACP* (New York: Berkley Publishing, 1962), 63.

    By espousing social and economic pressure instead of open violence to keep Oklahoma "100 percent American," the Klan reached out to a better class of society. Taft Welch, senior chairman of the Western National Bank of Tulsa, arrived in the city in 1923 as a boy of fifteen. He immediately joined the Klan, then opened a business selling members high-priced white robes. "It seemed like everybody had to join the Klan to do business," he remembers. "All the bank officers and businessmen and politicians were members. I saw what the Klan did and didn't like: they were against Catholics and Jews and blacks. They thought they were 100 percent American, and nobody else was."

    Perhaps because of the race riot, the Klan often selected smaller and less organized minorities to vent their hatred. Jews and Catholic immigrants (especially Italians) were the victims of choice. To this day Jewish Oklahomans—particularly those who have made money in the oil industry—go out of their way to keep a low profile. Julius Livingston arrived in 1909 and started Livingston Oil, one of the first Jewish oil companies in America. He became a friend of Jake Simmons's and bought a number of leases from him. In his office he still keeps a sixty-five-year-old death threat framed on the wall. It warned "Livingston and his Jew Crew" to get out of Tulsa or be killed.

    Livingston stood up to the Klan and nothing happened to him. A Jewish boardinghouse operator named Nate Hantaman was not so lucky. In August 1923 a mob whipped him and smashed his genitals "to a pulp." When Oklahoma governor John C. Walton was provided with evidence linking Klansmen in the police department with the incident, he declared war on the hooded order. The National Guard took over the offices of Tulsa's sheriff and police commissioner. A military court investigation into mob terrorism linked the Klan to twenty-five thousand whippings during the previous year. Walton tightened martial law and suspended the writ of habeas corpus in Tulsa, in violation of the state constitution. On September 2, in a comment which lionized him in the eyes of Oklahoma blacks, the governor publicly declared, "I feel just this way about the whipping crowd. The men who compose it are no better than murderers, burglars, and thugs. I don't care if you burst right into them with a double-barreled shot gun. I'll promise you a pardon in advance."

Although Walton was a rough-speaking populist hero (known by the working man as "Our Jack"), the Klan proved to be a more powerful enemy than he anticipated. His heavy-handed use of police powers gave lawmakers an excuse to make him the only Oklahoma governor ever removed from office. In the fall of 1923 state legislators impeached him, by a two-thirds majority, on charges of corruption and abuse of power.

Interviews with Julius Livingston, Donald Simmons, and Taft Welch; Neuringer, *Chronicles,* 153–79.

15. Interviews with Milton Nelson, Donald Simmons, and J. J. Simmons III.
16. Interviews with Ahmed Shadeed, Donald Simmons, and J. J. Simmons III.
17. Interviews with Donald Simmons and J. J. Simmons III.
18. Interviews with Congressman Sam B. Hall, Jr., John Montgomery, Donald Simmons, J. J. Simmons III, and John Simmons; *New York Times,* November 1, 1923.
19. Anthony Sampson, *The Seven Sisters* (New York: Viking, 1975), 90; Taft Welch, interviews.
20. Sampson, *Sisters,* 90; J. J. Simmons III, interviews.
21. Interviews with Edythe Rambo Fields, Mrs. E. C. Ryan, Donald Simmons, J. J. Simmons III; Nathaniel J. Washington, *Historical Development of the Negro in Oklahoma* (Tulsa: Dexter Publishing Co., 1948), 38.
22. Interviews with Edythe Rambo Fields, Donald Simmons, and J. J. Simmons III.
23. Interviews with Edythe Rambo Fields, Dr. John Montgomery, Mrs. E. C. Ryan, Donald Simmons, J. J. Simmons III, and Taft Welch.
24. Hughes, *Fight,* 11; interviews with Doris Montgomery, Dr. John Montgomery, J. J. Simmons III, and Taft Welch.
25. Interviews with Donald Simmons and Taft Welch.
26. Hughes, *Fight,* 53–55; Dallas Morning News, *Texas Almanac: 1988–1989* (Dallas: Dallas Morning News, 1987), 41.
27. Interviews with Donald Simmons.
28. Interviews with B. F. Phillips, Jr., Donald Simmons, and J. J. Simmons III.
29. Johnnie Mae Simmons, interview.

30. Interviews with B. F. Phillips, Jr., Donald Simmons, J. J. Simmons III, John Simmons, and Taft Welch.
31. Taft Welch, interviews.
32. Interviews with Congressman Sam B. Hall, Jr., B. F. Phillips, Jr., Donald Simmons, and J. J. Simmons III.
33. Interviews with Donald Simmons, J. J. Simmons III, and Taft Welch.
34. Interviews with Donald Simmons and J. J. Simmons III.
35. Louis R. Harlan and Raymond Smock, eds., *The Booker T. Washington Papers (1912–14),* Vol. 12 (Urbana, Chicago, and London: University of Illinois Press, 1982), 264.

Chapter 5 HOLDING OUT FOR DIGNITY

1. Interviews with Donald Simmons.
2. Taft State Hospital Records, Division of State Archives, Oklahoma City; *Black Dispatch,* December 1 and December 22, 1932, September 9, 1937.
3. *Black Dispatch,* September 9 and 16, 1937.
4. *Black Dispatch,* September 9 and 16, November 6, December 4, 1937.
5. *Black Dispatch,* September 9 and 16, 1937.
6. *Black Dispatch,* November 13, December 4, 1937.
7. *Black Dispatch,* November 13, 1937.
8. *Black Dispatch,* January 8, 1938.
9. *Black Dispatch,* November 13, December 4, 1937.
10. *Black Dispatch,* November 6 and 13, 1937.
11. *Black Dispatch,* January 15, 1938; *Muskogee Daily Phoenix,* January 20, 1938.
12. *Black Dispatch,* January 22, 1938.
13. *Muskogee Daily Phoenix,* January 20 and February 10, 1938; *Black Dispatch,* February 12 and February 26, 1938.
14. *Muskogee Daily Phoenix,* September 14, 15, 16, 1938.
15. Interviews with Donald Simmons and J. J. Simmons III; Hughes, *Fight for Freedom,* 55–56; Ellsworth, *Death,* 19; Teall, *Black History,* 220–21.
16. Interviews with J. J. Simmons III.
17. *Willa Eva Simmons* v. *Board of Education,* Muskogee, September 22, 1938. Amended Complaint (United States District Court for the Eastern District of Oklahoma, Doc. No. 4841), 10.
18. *Simmons* v. *Board of Education,* Original Bill of Complaint, September 14, 1938, and Amended Complaint September 22, 1938, 11; *Muskogee Daily Phoenix,* September 16, 1938.

19. "*Simmons* v. *Board of Education,*" *Muskogee Daily Phoenix,* September 16 and October 4, 1938.
20. *Muskogee Daily Phoenix,* October 1, 1938.
21. Ibid.
22. Ibid.
23. *Muskogee Daily Phoenix,* October 3, 1938.
24. *Muskogee Daily Phoenix,* October 4, 1938; *Simmons* v. *Board of Education,* October 12, 1938, Findings of Fact.
25. *Muskogee Daily Phoenix,* October 4, 1938; *Simmons* v. *Board of Education,* September 14, 1938, Amended Complaint and October 12, 1938, Findings of Fact.
26. *Muskogee Daily Phoenix,* September 16 and October 4, 1938.
27. *Muskogee Daily Phoenix,* October 4, 1938.
28. Donald Simmons, interviews; *Muskogee Daily Phoenix,* October 13, 1938, and January 8, 1939.
29. Interviews with Donald Simmons and J. J. Simmons III.
30. Donald Simmons, interviews.
31. Ibid.
32. Ibid.
33. Ibid.
34. *Muskogee Daily Phoenix,* October 13, 1989; *Simmons* v. *Board of Education,* October 12, 1938, Findings of Fact.
35. *Black Dispatch,* October 29, 1938.
36. Ibid.
37. *Black Dispatch,* December 17, 1938; *Muskogee Daily Phoenix,* January 4 and January 5, 1939; *Simmons* v. *Board of Education,* January 4, 1939, Narrative Statement of Evidence.
38. *New York Times,* May 23, 1939; *Lane* v. *Wilson,* 307 U.S. 268 (1939); *Plessy* v. *Ferguson,* 163 U.S. 537 (1896); *Sipuel* v. *Board of Regents,* 332 U.S. 631 (1948); *McLaurin* v. *Oklahoma State Regents,* 339 U.S. 637; *Brown* v. *Board of Education of Topeka,* 347 U.S. 483 (1954); Teall, *Black History,* 223–24; Richard Hofstadter, ed., *The United States 1916–1970: A World Power.* 4th ed. (Englewood Cliffs, New Jersey: Prentice-Hall, 1976), 338.
39. *Simmons* v. *Board of Education,* 306 U.S. 617 (1938); February 27, 1939, Supreme Court decree; *Muskogee Daily Phoenix,* March 17, 1939.

Chapter 6: THE BUILDING BLOCKS OF PRIDE

1. Donald Simmons, interview.
2. Donald Simmons, interview.
3. Interviews with Donald Simmons and J. J. Simmons III.
4. Interviews with Donald Simmons and J. J. Simmons III.
5. J. J. Simmons III, interviews.
6. J. J. Simmons III, interviews.
7. Interviews with Blanche Jamierson and Johnnie Mae Simmons.

   Although "Mother Eva" was not Blanche's real mother she treated her like her own daughter. Eva encouraged her husband to support her, both financially and emotionally. "I just felt that I was always her girl—I could come to her with anything," Blanche says. "She taught me to sew, and she always sent me clothes in the fall and at Christmas time. Mother Eva put every stitch of clothes on me that I ever wore."
8. Johnnie Mae Simmons, interviews.
9. J. J. Simmons III, interviews.
10. Interviews with Donald Simmons.
11. Interviews with Donald Simmons, J. J. Simmons III, and Johnnie Mae Simmons.
12. J. J. Simmons III, interviews.
13. Interviews with Donald Simmons, J. J. Simmons III, and Johnnie Mae Simmons.
14. Interviews with Doris Montgomery, Donald Simmons, J. J. Simmons III, and Johnnie Mae Simmons.
15. The Reverend E. W. Dawkins, interview.
16. Interviews with the Reverend E. W. Dawkins and Donald Simmons.
17. The Reverend E. W. Dawkins, interview.
18. Ibid.

    Among the psalms Simmons frequently repeated were the Twenty-third ("The Lord is my shepherd; I shall not want . . .") and the 133rd ("Behold how good and how pleasant it is for brethren to dwell together in unity . . ."). At home the family began each meal with any of a number of familiar prayers, an occasion Simmons frequently used to make a speech, adding whatever personalized messages were on his mind. J. J. III says that his father "talked to God like, 'This is my God up there.' That's why," the son observes, "he was so disappointed when he was near death. He just couldn't believe that God was doing this to him because this was the God he'd been talking to all these years, letting him down." Later in life, recalls granddaughter Donna Simmons, Jake Jr. was more apt to repeat a regular benediction, "In all thy ways acknowledge Him, and He will direct thy life." Donna still says this prayer before she eats. "When you really think about it," she says, "that is sort of his life. Just that saying."

19. Interviews with the Reverend E. W. Dawkins, Donald Simmons, and Johnnie Mae Simmons.
20. Interviews with Donald Simmons and J. J. Simmons III.
21. Almetta Carter, interview.
22. Donald Simmons, interview.
23. *Muskogee Daily Phoenix*, April 2, 1946; Teall, *Black History*, 225–32; interviews with Napoleon Davics, Ed Edmondson, Jr., and Donald Simmons.
24. *Muskogee Daily Phoenix*, April 2, 1946.
25. *Muskogee Daily Phoenix*, April 3, 1946.
26. Interviews with Donald Simmons and James Stewart; *Oklahoma Eagle*, June 15, 1946; *Black Dispatch*, June 15, 1946.
27. Oklahoma Department of Libraries, *Directory of Oklahoma* (Oklahoma City: Oklahoma Department of Libraries, 1981), 734–35.
28. Interviews with Congressman Ed Edmondson, Jr., and Donald Simmons; *Directory of Oklahoma, 1985*, 577.
29. Interviews with Ed Edmondson, Jr., and John Hannah; *Muskogee Daily Phoenix*, July 25, 1948.
30. *Muskogee Daily Phoenix*, July 28, 1958.
31. Hughes, *Fight for Freedom*, 99–100; 136–39; *Black Dispatch*, March 30, 1946.

In 1946 black student Ada Sipuel and the NAACP sued the University of Oklahoma for refusing her admittance to its law school. There were no law schools for blacks in Oklahoma and the state NAACP waged a three-year legal battle. Finally in 1949 the Supreme Court affirmed her right to study there and forced the state to revise its segregation laws.

The University of Oklahoma tried to avoid the impact of the ruling by making life tough on is few black students. When G. W. McLaurin was admitted to the graduate school of education he was made to sit in an anteroom apart from the class, study at a special table in the library, and eat in the cafeteria, in a specified seat, at specified hours. In 1950, the Supreme Court declared that this policy violated McLaurin's right to equal protection under the Fourteenth Amendment, and ordered the university to end its segregation of McLaurin.

32. *Jake Simmons, Jr.*, v. *Board of Education of Independent School District No. 20 of the City of Muskogee, and Board of County Commissioners of Muskogee County, Oklahoma*, February 18, 1949, Complaint (State of Oklahoma

District Court, Muskogee County, Doc. No. 32899); *Muskogee Daily Phoenix,* March 22, 1949, and March 24, 1949; *Tulsa World,* March 23, 1949.

33. *Simmons* v. *Board of Education,* Temporary Restraining Order, February 21, 1949; *Simmons* v. *Board of Education,* Answer of Defendant, Board of Education, March 21, 1949.
34. *Directory of Oklahoma, 1985,* 575; *Black Dispatch,* July 1, 1950.
35. William H. Murray, *The Negro's Place in Call of Race* (Tishomingo, Oklahoma: William H. Murray, 1948), 22–23.
36. *Oklahoma Eagle,* July 20, 1950.
37. *Oklahoma Eagle,* June 22, 1950.
38. *Black Dispatch,* July 1 and July 8, 1950.
39. Donald Simmons, interviews; *Black Dispatch,* July 8 and September 16, 1950.
40. *Black Dispatch,* July 29, 1950.
41. Interviews with B. F. Rummerfield and J. J. Simmons III; résumé of J. J. Simmons III.
42. In 1984 the AAPG awarded J. J. Simmons III its highest honor, The Public Service Award. *Oklahoma Eagle,* September 9, 1954; *Black Dispatch,* September 18, 1954; interviews with B. F. Rummerfield and J. J. Simmons III.
43. Interviews with Donald Simmons and J. J. Simmons III.
44. Ibid.
45. Ibid.

Rufus Dunlop, a white cable tool driller and tool dresser who worked on a number of Simmons's wells during this period recalls that Jake Jr. "treated me better than anyone I'd ever been around before." Dunlop, the son of a tenant farmer, has lived his whole life in eastern Oklahoma. Simmons took a liking to the good-natured, powerfully built young worker. As he often did with people he met, Simmons tried to entice Dunlop to aspire to a more glorious career. "He wanted me to go to Dallas to be a prizefighter," recalls Dunlop, with a broad Oklahoma grin. "He knew a trainer down there and said he'd take care of my family while I was gone. He was going to promote me for a percentage." Dunlop passed on the offer, though he has no doubt it was a serious one. He has similarly fond memories of J.J. III. It was the extra little things, like buying their lunch at the local barbecue or supplying them with free cigars, which won the workers' allegiance. Dunlop still remembers J.J. promising to buy him a new suit of clothes if the well he was drilling "came in," and how pleased he was when they struck oil.

46. J. J. Simmons III, interviews.
47. Ibid.
48. Ibid.

Chapter 7 THE ONE THAT GOT AWAY

1. Dr. Jesse Chandler, interview.
2. Robert D. Thomas et al., *Area Handbook for Liberia* (Washington, D.C.: The American University, 1972), 1–22; Philip M. Allen and Aaron Segal, *The Traveler's Africa: A Guide to the Entire Continent* (New York: Hopkinson and Black, 1973), 541; "Made in America," *U.S. News & World Report,* August 6, 1954, 54.
3. Thomas, *Liberia,* xix–xxvii; "Highlights from the Inauguration," *Liberia Today,* July 1952, 3–7; "Honoring 'President' Tubman," *Nation,* January 26, 1952, 72.
4. Interviews with Jewel Hines, Donald Simmons, and J. J. Simmons III; *Black Dispatch,* January 5 and February 2, 1954; *Liberia Today,* 3–7; "Liberia Asks 50 Percent of Profits," *Business Week,* July 5, 1952, 80; Thomas, *Liberia,* xxiv, 268–70.
5. Interviews with Donald Simmons and J. J. Simmons III; Thomas, *Liberia,* 66–67; *Oklahoma Eagle,* September 15, 1955.
6. J. J. Simmons III, interviews; *Oklahoma Eagle,* January 8, 1953.
7. *Muskogee Daily Phoenix,* April 6, 1954; *Oklahoma Eagle,* April 1, 1954.
8. *Muskogee Daily Phoenix,* April 6, 1954.
9. *Muskogee Daily Phoenix,* April 7, 1954.
10. *Muskogee Daily Phoenix,* April 7 and 9, 1954.
11. *Black Dispatch,* March 14, 1958; *Muskogee Daily Phoenix,* April 6, 1954.
12. Interviews with Dr. Jesse Chandler and Donald Simmons.
13. Dr. Jesse Chandler, interview.
14. Interviews with Dr. Jesse Chandler and Emery Jennings.
15. *Black Dispatch,* May 29, June 5, and June 12, 1954; "How Did Your Congressman Vote? NAACP Legislative Scoreboard for the 85th Congress," September 1958, Papers of the NAACP, Library of Congress.
16. Joseph Kraft, " 'King' of the U.S. Senate," *Saturday Evening Post,* January 11, 1936, 26–27; Daniel Seligman, "Senator Bob Kerr, the Oklahoma Gusher," *Fortune,* March 1959, 138–88; Otis Sullivant, "Rich Man's Race," *Nation,* June 26, 1954, 542–43; "Oklahoma's Kerr—the Man Who Really Runs the U.S. Senate," *Newsweek,* August 6, 1962, 15–17; "Uncrowned King of the Senate," *New Republic,* January 12,

1963; Kerr-McGee, *1965 Annual Report;* interviews with Emery Jennings and B. F. Rummerfield.

17. *Black Dispatch,* June 12, 1954.
18. *Black Dispatch,* May 29, 1954.
19. Ibid.
20. Program for Roscoe Dunjee's testimonial dinner, June 27, 1952; Papers of the NAACP; "Dunjee Pioneered for Negro Rights," *Pittsburgh Courier,* July 9, 1955; Donald Simmons, interviews.
21. Interviews with Mrs. E. L. Goodwin and Emery Jennings.
22. *Oklahoma Eagle,* June 3, 1954.
23. Interviews with Joseph E. Howell and Melvin Porter.
24. Donald Simmons, interviews.
25. *Black Dispatch,* July 3, 1954; *Oklahoma Eagle,* July 1, 1954; Emery Jennings, interview.
26. Emery Jennings, interview.
27. *Oklahoma Eagle,* July 1, 1954.
28. Ibid.
29. *Oklahoma Eagle,* July 1 and July 21, 1954; *Black Dispatch,* July 10 and July 17, 1954.
30. Interviews with Donald Simmons and J. J. Simmons III.
31. Donald Simmons, interviews.
32. J. J. Simmons III, interview.
33. Ibid.
34. Ibid.
35. Interviews with J. J. Simmons III, Chuck Snacke, and Clyde Wheeler; Sunray Mid-Continent Co., *1956 Annual Report.*
36. Interviews with Donald Simmons and J. J. Simmons III.
37. Donna McFarland, telephone interview.
38. Sunray DX Co., *Sunray News,* May–June 1966, July–August 1968; *1967 Annual Report;* interviews with Donald Simmons and J. J. Simmons III.
39. Robert A. Smith, *The Liberia Annual Economic Review* (Monrovia, Liberia: Providence Publications; New York: Liberia Investment Services Corporation, 1972), 19–24; *Liberia: Geographical Mosaics of the Land and the People* (Monrovia, Liberia: Liberian Ministry of Information, Cultural Affairs and Tourism, 1979), 4–8; Roberts, *Liberia,* 268–70; J. J. Simmons III, interview.

Chapter 8 BACK TO AFRICA

1. W. H. Thompson, interview.
2. Bartlesville Chamber of Commerce, *Bartlesville,* promotional brochure, 1986.
3. Employees say that during the past few decades the company has made a conscious effort to hire minorities, especially Native Americans, but the effect of this effort was far from evident in the racial disposition of those I saw near the top of the corporate ladder.

   William C. Wertz, ed., *Phillips—The First Sixty-Six Years* (Bartlesville, Oklahoma: Phillips Petroleum, 1983), 13–21; Robert Gregory, *Oil in Oklahoma* (Muskogee, Oklahoma: Leake Industries, 1976), 24; Sarah Peterson, "Bartlesville, a 'Company Town' and Proud of It," *U.S. News & World Report,* October 22, 1984, 59; Chamber, *Bartlesville;* interviews with Sally Dennison, Len M. Rickards, Owen Thomas, and Edwin Van den Bark.
4. Sally Dennison, interview.
5. Anthony Sampson, *The Seven Sisters* (New York: Viking, 1975), 78–79.
6. Phillips Petroleum corporate documents, Bartlesville, Oklahoma; Owen Thomas, interview.
7. J. J. Simmons III, interviews.
8. Interviews with Ed Goodwin, Donald Simmons, and J. J. Simmons III.
9. Interviews with Roy Gootenberg, Maurice Keville, and Donald Simmons, *Oklahoma Eagle,* October 17, 1963.
10. Donald Simmons, interview.
11. Interviews with Roy Gootenberg, Donald Simmons, and J. J. Simmons III; Phillips Petroleum corporate documents.
12. "Americans of African descent," he told the audience, had managed to make themselves "indispensable in the agricultural life of the southland." Like his Tuskegee mentor, the Oklahoman followed this statement with a barrage of statistics, noting, among other things, that blacks owned approximately one-fifth of the agricultural land in the U.S., as well as major insurance companies and banks with assets exceeding $1 billion. He predicted that a black would some day be president of the American Stock Exchange, and closed his speech on the following note of optimism: "And I say this with a deep sense of humility and also a great sense of pride saying that America offers the opportunity. It's the country of multiplicities of races. And if you are possessed with grim determination, with bulldog tenacity, you can certainly reach your goal."

    Oklahoma Eagle, October 17, 1963; interviews with Roy Gootenberg and Donald Simmons; Voice of America Report on Jake Simmons's speech in the Dar es Salaam broadcast, December 4, 1963; *New York Times,* May 13, 1964.

13. Audiotape, Voice of America, December 4, 1963.
14. Interviews with Roy Gootenberg and Donald Simmons.
15. Interviews with Yusuff Maitama-Sule and Donald Simmons.
16. Phillips Petroleum, *1971 Annual Report,* 24; *1985 Annual Report,* 51; A. Alvarez, "Offshore," *New Yorker,* January 20, 1986, 49; Donald Simmons, interviews.
17. Edwin Van den Bark, interview.
18. Sampson, *Seven Sisters,* 176–80; *Daily Times, Nigeria Year Book: 1966* (Lagos, Nigeria: Time Press, 1966) 271–73; "Additional Operators Will Explore for Oil in Nigeria," *World Oil,* June 1962, 124; Richard Graham, "Nigeria: Oil Country Now, *International Commerce,* July 20, 1964, 11.

    Mattei first set the oil industry on edge in 1957, by giving the Shah of Iran a bigger share of the profits and a more active participatory role than any oil company had ever granted to a country.
19. "Phillips, Agrip Rush Plans to Produce First Nigerian Oil," *Oil and Gas Journal,* February 13, 1967, 93; Phillips Petroleum, corporate reports; interviews with Donald Simmons and Edwin Van den Bark.

    Mattei died before the terms of the deal were concluded. The iconoclast's death was shrouded in controversy; even today a rumor persists that the fatal crash of Mattei's private plane was engineered by sinister rivals within the petroleum industry.
20. *Oil and Gas Journal,* February 13, 1967, 93; interview with L. M. Rickards, Donald Simmons, Owen Thomas, and Edwin Van den Bark.

    The Nigerian government's end of the deal was a 50-percent tax on all profits, plus a cost-free option to buy a large stake in the project if it chose to.
21. U.S. tax regulations regarded stock interests in foreign subsidiaries as passive investments. This meant that unless Phillips got a "working interest," the hefty income taxes it paid to the Nigerian government would not be deductible from the company's U.S. income tax.

    Interviews with Donald Simmons, Owen Thomas, and Edwin Van den Bark.
22. Interviews with L. M. Rickards, Donald Simmons, Owen Thomas, and Edwin Van den Bark.
23. "How They Won the West—and More," *Business Week,* January 28, 1967, 178.
24. Interviews with Sally Dennison, Donald Simmons, Owen Thomas, and Edwin Van den Bark.

25. Interviews with Billy Adams and Roy Gootenberg.
26. Wertz, *Phillips,* 138; "Phillips Petroleum: Laying the Groundwork," *Forbes,* February 1, 1965, 22; interviews with Bill Adams, L. M. Rickards, and Donald Simmons.
27. Interviews with Sylvan Amegashie, Horace G. Dawson, Jr., Henry Aliva Johnson, Yusuff Maitama-Sule, Donald Simmons, Owen Thomas, and Edwin Van den Bark.
28. Interviews with Donald Simmons and J. J. Simmons III; "Cheering Harlem Rally Hears Malcolm X Rip U.S. Racism," *The Militant,* June 15, 1964; 1; Alex Haley and Malcolm X, *The Autobiography of Malcolm X* (New York: Grove Press, 1964), 349–52, 431.
29. Haley, *Autobiography,* 349–52, 431; *New York Times,* May 13 and May 18, 1964; Donald Simmons, interviews.

    From Malcolm X as well as Minister Sule and other Muslim officials in Nigeria, Simmons developed a new respect for Islam. Late in his life, when his twenty-four-year-old nephew "Junior" changed his name to Ahmed Shadeed and became an observant Muslim, Simmons confronted him about the decision.

    "I hear you're a Muslim?" he asked.

    "Yes, sir," Ahmed replied.

    "You changed your name?"

    "Yes, sir."

    "You didn't disown your family?"

    "No, sir."

    "Because you will *never* disown this family."

    "Yes, sir."

    "That's good," concluded Simmons. "All the Muslims I know have got their lives together."
30. *New York Times,* August 28, 1985; *International Commerce,* July 20, 1964, 14; Harold D. Nelson, ed., *Nigeria, A Country Study* (Washington, D.C.: The American University, 1982), 295.
31. Interviews with Yusuff Maitama-Sule, Owen Thomas, and Edwin Van den Bark.

    Owen Thomas questions Van den Bark's recollection, and provided me with a detailed account of how the approval was achieved. Thomas says he arrived in Lagos on February 14, 1965, expecting to stay one week. He met Simmons right away and the pair planned a series of meetings. The negotiations, some of which Simmons attended, lasted more than six weeks. Thomas had hoped for help from AGIP. "When I got there," he explains, "they only had their field staff there. No one from headquarters. So I fired an angry letter to Milan for help. We had to take the initiative. I said, 'I'm not coming back till we get this thing approved.' We weren't even sure of getting the stock interest. If we hadn't gotten the deal approved we would have been out, and AGIP would have owed us one. But when and where we would have gotten it back we didn't know."

In the end, according to Thomas, it was not Simmons's ability as an intermediary, but an offer to toss in a bundle of cash which "clinched" the deal. "We tried to explain it to Sule," says Thomas. "We'd drawn pie diagrams and everything to explain the difference between a stock interest and a working interest, but he didn't understand. He wasn't sure he wanted to do it. I was getting sick of it. Our lawyer came back [from Bartlesville] and got approval for $1 million. When we added $1 million in cash to the Nigerian government Sule said, 'Now I understand.' "

That Phillips would have had this kind of cash on hand for payment to a foreign government is not unusual in the international oil business. Without it the company would never have been able to compete with other multinationals at the time. During the mid-1970s the Securities and Exchange Commission investigated the oil giants. It forced Gulf, in its 1975 proxy statement, to divulge that its chairman had paid $4 million in bribes to the reigning party in Korea, and $350,000 to the ruler of Bolivia. Ashland Oil admitted to making nearly $500,000 in questionable payments to government officials and consultants in Nigeria, Gabon, the Dominican Republic, and Libya. Investigations by government agencies also revealed that Phillips Petroleum maintained a secret $3 million slush fund in Swiss bank accounts. According to an action brought by the SEC in 1975, the lion's share of the $3 million "was distributed overseas in cash," and the company used "international couriers, code names, misleading entries, and false invoices and billings to conceal the overseas funds from the Internal Revenue Service." An unresolved lawsuit filed later by the Justice Department contended that the Swiss accounts accumulated as much as $42.85 million in 1963 and 1964. Prosecution by the SEC and the Justice Department regarding Phillips Petroleum's international slush fund was cut short by a blanket plea-bargaining arrangement relating to an illegal $100,000 contribution by the company's chairman to President Nixon in 1972. A report on the contributions and Swiss funds commissioned by Phillips stated that none of the money paid to foreign associates through a Swiss corporation was used to pay "any bribe, commission, contribution, or other payment to any government official, politician, or political party."

Owen Thomas does not recall ever actually paying the $1 million to the Nigerian government. After a series of disputed elections in 1965,

Sule was deposed as oil minister. His permanent secretary Alhaji Musa Daggash filled the power void during the ensuing years of civil strife. Daggash, who, like Sule, became close friends with Simmons and used him as an intermediary, concluded the AGIP deal.

Phillips Petroleum documents; Sampson, *Seven Sisters,* 247; *New York Times,* March 7, 1975, August 24, 1975, section 3, September 27, 1975, September 3, 1976, May 18, 1977, July 6, 1977, November 23, 1977.

32. Phillips Petroleum documents; interviews with Donald Simmons and Owen Thomas.
33. Executive vice president Rickards, when asked about any old datebooks he might have, commented, "I don't save diaries. I used to have a stack of these annual diaries, but anytime you make a margin note even on a piece of correspondence that becomes an original and [with] all sorts of litigation that we've had—takeover attempts and everything—every original document is put on hold and you fill up all your storage records with that stuff. So I think most everybody around here has destroyed everything that they ever kept."

    L. M. Rickards, interview.
34. Phillips Petroleum documents.
35. Interviews with L. M. Rickards, Donald Simmons, and Edwin Van den Bark; Phillips Petroleum corporate reports and documents.
36. Phillips Petroleum, *Information for Analysts,* First Quarter, 1986.
37. Interviews with L. M. Rickards, Donald Simmons, Owen Thomas, and Edwin Van den Bark; *New York Times,* April 27, 1967.
38. Phillips Petroleum documents; interviews with L. M. Rickards, Owen Thomas, and Edwin Van den Bark.
39. Phillips Petroleum, *Annual Reports, 1965–1974;* Phillips Petroleum, *Information for Analyst Reports,* 1978–1986; *Oil and Gas Journal,* February 13, 1967, 93; Nelson, *Nigeria,* 52–60.
40. L. M. Rickards, interview; Phillips Petroleum documents.
41. Interviews with Donald Simmons and Edwin Van den Bark.
42. Kwame Nkrumah was one of the fathers of African nationalism. After studying in America (at Lincoln University and the University of Pennsylvania) and England, he returned to Ghana with a passion for independence. His emotional speeches ignited the imagination of the average Ghanaian, and he forged a populist party which swept the British from power well before they wanted to leave. Once in control he built a highly idealistic revolutionary government. "Any compromise over principle," he wrote, "is the same as an abandonment of it."

    Nkrumah founded the Organization of African Unity. He encouraged his neighbors to reject what he called "neocolonialism," and promoted the establishment of one unified Pan-African state. He agitated for true independence and didn't want to wait for it; he reportedly provided training and arms for guerrillas from other African nations. His government was so virulently anticolonialist that many thought the American CIA later had a hand in his overthrow.

On an economic level Nkrumah was eager to press for full development of an industrial infrastructure. He built an enormous port city to stimulate commerce, and oversaw the construction of one of the world's largest hydroelectric plants. His enthusiasm knew no bounds. Since Ghana was the world's largest cocoa producer he attempted to dictate the commodity's price by building gargantuan silos to hoard all the cocoa in the world and allow small agricultural nations like his to deal with international capitalists from a position of strength. In modern Africa Nkrumah is looked back upon as a daring visionary twenty years ahead of his time, as a Marxist fanatic, or as a misguided idealist—depending on whom one speaks to.

Donald Simmons, interviews; Shirley Graham Du Bois, "What Happened in Ghana? The Inside Story," *Freedomways,* Summer 1964, 201–23; Kwame Nkrumah, "The Merchants of Neo-Colonialism," *Freedomways,* Second Quarter, 1966, 139–58; Horace Mann Bond, *Education for Freedom* (Pennsylvania: Lincoln University, 1976), 505–508; Freedom Fighters Edition, *Axioms of Kwame Nkrumah* (London: Panaf Books, 1969), 54; Irving Kaplan et al. *Area Handbook for Ghana* (Washington, D.C.: The American University, 1971), 70–86.

43. Interviews with Sylvan Amegashie and Donald Simmons.
44. Interviews with Zvi Alexander, Sylvan Amegashie, Marvin Billet, Donald Simmons, and Loring B. Sneden; *Ghanaian Times,* December 17, 1968.
45. Interviews with Sylvan Amegashie and Donald Simmons.
46. Donald Simmons, interviews.
47. National Petroleum Council, *Memorial Resolution to J. J. Simmons, Jr.,* April 16, 1981.
48. Nelson, *Nigeria,* 51–60; Donald Simmons, interviews.
49. An eight-member "Liberation Council" of military and police officers took over the government. Kwame Nkrumah became an exile, spending the last six years of his life in Guinea. His supporters believed the action was brought about with the complicity of the CIA and spies for other Western countries. The widow of W. E. B. Du Bois, Shirley Graham Du Bois, was director of Ghana television under Nkrumah. Writing in the

American magazine *Freedomways* shortly after the coup, she asserted that there was no populist support for the uprising, and that the troops participated in the action only because they had been told that Nkrumah had left the country for Red China with a fortune in gold, that he was preparing to send his army to fight for the North in Vietnam, and that the Russians were about to take control of the country, having arrived secretly at Flagstaff by way of a secret tunnel from the airport. "Any resemblance to a revolt on the part of the people of Ghana," she wrote, "had to be manipulated and fabricated by skillful, directed intelligence from the outside."

Many other Ghanaians, however, believed the uprising enjoyed widespread support, inspired by Nkrumah's tyranny. In one of the few contemporary political accounts to be published in Ghana, distinguished Ghanaian journalist Mike Oquaye noted that despite the essential economic development accomplished under Nkrumah's reign and the fact that he successfully "ushered in a rebirth of the African personality and inspired the rest of the continent to attain independence," an abuse of power proved to be the revolutionary's downfall. "Nkrumah's overreaching expenditure on the continent depleted resources at home," Oquaye wrote, "and in the midst of poverty and want his 'socialist boys' lived privately like lords. . . . Opponents of the regime—political ones at first and subsequently competitors in private rivalries—were arbitrarily detained. Every conceivable freedom—academic, press, speech, and association—was strangled. Judges were arbitrarily dismissed, a one-party state was declared . . . and Parliament was shamefully reduced to a mere rubber stamp of Nkrumah's."

Mike Oquaye, *Politics in Ghana* (Accra, Ghana: Tornado Publications, 1980), 2–3; Christine Johnson, "Letter on the Ghana Coup," *Freedomways,* Summer 1966, 152–58; Patrick Smith, "Confessions of the CIA," *New African,* December 1985, 8; Du Bois, *Freedomways,* 201–23; *Ghana Area Handbook,* 5–6, 80–86; *Axioms of Kwame Nkrumah,* 67.

50. Oquaye, *Politics,* 3; Kaplan, *Ghana,* 203–204; Sylvan Amegashie, interview.
51. Interviews with Loring Sneden and W. H. Thompson.
52. Ibid.
53. Loring Sneden, interview.
54. Interviews with Sylvan Amegashie, Loring Sneden, and W. H. Thompson; *Ghanaian Times,* December 20, 1968; *Oil and Gas Journal,* December 15, 1969, 50, and July 13, 1970, 38.
55. Interviews with Sylvan Amegashie and Loring Sneden.

56. Sylvan Amegashie, interviews.
57. Interviews with Sylvan Amegashie, Anum Barnor, Robert E. Lee, Loring Sneden, and Edith Tamakloe.
58. Haley, *Malcolm X,* 352–53; *Oil and Gas Journal,* July 13, 1970, 38; interviews with Zvi Alexander, Sylvan Amegashie, Marvin Billet, Loring Sneden, and W. H. Thompson.
59. The Ghanaian government also received various payments along the way, most of them adhering to a standardized schedule which applied to other oil companies as well. There were varying introduction bonuses, then a charge of $7,000 per parcel for a seismic option. Concession holders had to make annual rental payments rising from twenty-five dollars a square mile the first year to fifty dollars a mile the third (for the Signal group this would have cost about $40,000 the first year). Because the territory was unproven and risky, the Ghanaian government had to tie most of its income to a 50-percent tax it would receive on the profits of all successful operations. Regardless of the outcome, the ministry's sparse pool of seismographic data would benefit from the tens of millions of dollars in research and exploration the multinationals committed themselves to. The most advantageous aspect of the Ghanaian petroleum code was its stringent performance obligations. In order to retain their concessions, the oil companies were required to begin drilling exploratory wells within twenty-four months, to use modern equipment, to provide the oil ministry with prospecting reports at regular intervals, and to bore to an aggregate depth of no less than 12,000 feet (an operation which could not be done for less than a few millions dollars). Within eighteen months each company had to submit a plan to train Ghanaians for skilled oil-related positions. After the twenty-four-month exploration period the concessionaire had a choice of paying for an operational license or giving up its property. If the company surrendered the area, the Ghanaians would be able to use its expensive geological data to help resell the concession to other companies. On top of all this, if any company later offered better oil prospecting terms to any other country, Ghana's officials would be able to renegotiate its contracts on the same terms.

    Interviews with Zvi Alexander, Sylvan Amegashie, and Loring Sneden; *Ghanaian Times,* December 20, 1968.
60. Loring Sneden, interview; Kaplan, *Ghana,* 295.
61. Interviews with Loring Sneden and W. H. Thompson.
62. Interviews with Zvi Alexander and Sylvan Amegashie; *Ghanaian Times,* December 20, 1968; *Oil and Gas Journal,* July 13, 1970, 38, and January 1, 1969.

63. Sampson, *Seven Sisters*, 275; *Oil and Gas Journal*, July 13, 1970, 38, and May 31, 1971, 25; interviews with Loring Sneden and W. H. Thompson.
64. I was unable to trace all of Simmons's dealings in Ghana without the help of the government records, which were never made available to me. The current revolutionary government of Ghana is acutely xenophobic, especially toward Americans. A few years ago authorities discovered a network of CIA spies working out of Accra, whom they suspected of being involved in repeated coup attempts against Flight Lieutenant J. J Rawlings. (Known to his people as "Jerry," Rawlings is a tremendously dynamic young officer with considerable populist support who has been the country's sole ruler since 1979.) Although my request for information had the backing of a number of important Ghanaians—who told me that their government kept excellent records—after three months the current secretary for fuel and power, G. P. N. Grimah, informed me that his ministry was unable to trace any documents relating to Jake Simmons, Jr. Suspicious that he, too, might be a spy, the authorities interrogated a friend of mine there who attempted to assist in expediting my request.

    Interviews with Michael Akotu-Sasu, Sylvan Amegashie, and Bill St. John; *New African*, December 1985, 7–9; letter to author from Secretary for Fuel and Power G. P. N. Grimah, January 12, 1987.
65. Interviews with Michael Akotu-Sasu, Sylvan Amegashie, Bill St. John, Wayne Swearingen, and Jack Zarrow.

Chapter 9: A SEAT AT THE TABLE

1. Clara Luper, interview.
2. Teall, *Black History*, 241–57; Hofstadter, *World Power*, 668; Benjamin Quarles, *The Negro in the Making of America* (New York: Collier, 1971), 252–53; interviews with Clara Luper and Melvin Porter.
3. Melvin Porter, interview; Teall, *Black History*, 228; *Black Dispatch*, November 6, 1964.
4. Martin Luther King explained the effectiveness of nonviolent resistance as follows: "Nonviolence can touch men where the law cannot reach them . . . by appealing to the consciences of the great decent majority who through blindness, fear, pride, or irrationality have allowed their consciences to sleep."

    David L. Lewis, *King: A Biography* (Urbana, Illinois: University of Illinois Press, 1978), 85, 113.

5. Clara Luper, *Behold the Walls* (Oklahoma City: Wire Publishing, 1979), 91; Bill Harmon, "When Men Forget Color," *Oklahoma Orbit,* August 25, 1963, 7.
6. *Black Dispatch,* July 8, 1960.
7. Harmon, *Orbit,* 7.
8. Program Notes for NAACP Oklahoma Conference Annual Meeting, November 13, 1963; J. J. Simmons, Jr., letter to John Morsell, NAACP assistant executive director, October 31, 1966; Box 129, Papers of the NAACP, Manuscripts Division, Library of Congress.
9. Interviews with Donald Simmons and J. J. Simmons III.
10. Hofstadter, *World Power,* 683–89.
11. *Daily Oklahoman,* August 17, 1967.
12. Interviews with Almetta Carter, John Hannah, John Montgomery, and Donald Simmons.
13. Almetta Carter, interview.
14. James Barker, interview.
15. Jones's own political ambitions were cut short by a failed senatorial bid in 1986. His role as chief of staff in Johnson's White House was so high a position at so young an age that Jones now acridly refers to himself as "a has-been at twenty-nine."
16. James R. Jones, interview.
17. *Tulsa World,* November 10, 1967.
18. Interviews with Dr. Robert E. Lee and Donald Simmons; *Business Week,* August 9, 1969, 70–72.

    The article said that "Attracting Simmons was, in itself, a significant watershed for HCC." One of the purposes of "black capitalism" was to bring back to the ghetto members of the Negro elite who had "made it" in the white world.
19. Donald Simmons, interviews.
20. Interviews with the Reverend E. W. Dawkins, Bill Friman, and James Leake.
21. Interviews with Horace G. Dawson, Jr., Doris Montgomery, John Montgomery, Loring Sneden, Wayne Swearingen, Owen Thomas, and W. H. Thompson; Jake Simmons, Jr., letter to sons, January 17, 1972.
22. "He Went a Little Further," undated essay by Jake Simmons, Jr.
23. The Reverend E. W. Dawkins, interview; *The A.M.E. Review,* April–June 1964, 12.
24. *Daily Oklahoman,* September 10, 1968; *Black Dispatch,* September 11, 1969.
25. Interviews with George Nigh, James Leake, Donald Simmons, and J. J. Simmons III; *Daily Oklahoman,* March 10, 1966.

26. Interviews with John Hannah, Taft Welch, and Wayne Swearingen.
27. Interviews with Doris Montgomery, Mrs. M. L. Sanders, and Donna Simmons.
28. Interviews with B. F. Rummerfield, Donald Simmons, Wayne Swearingen, and W. H. Thompson.
29. *Daily Oklahoman,* July 30, 1971; National Petroleum Council, Memorial Resolution for J. J. Simmons, Jr., April 16, 1981; interviews with Donald Simmons and J. J. Simmons III.
30. Interviews with Donald Simmons and Jack Zarrow; Phillips Petroleum corporate documents; *Daily Oklahoman,* September 26, 1973.
31. *Daily Oklahoman,* July 2, 1972, and March 23, 1979; letters from J. J. Simmons III to Blanche Jamierson, December 31, 1975, December 15, 1976, and April 23, 1979.
32. Interviews with Doris Montgomery, Donald Simmons, W. H. Thompson, and Henry Zarrow; *Muskogee Phoenix,* March 26, 1981.
33. Interviews with Michael Akotu-Sasu, Bill St. John, Wayne Swearingen, and Jack Zarrow; *Ghanaian Times,* November 25, 1978; *Daily Graphic,* November 25, 1978.
34. Interviews with Zvi Alexander, Michael Akotu-Sasu, Sylvan Amegashie, Bill St. John, Donald Simmons, J. J. Simmons III, Wayne Swearingen, and Jack Zarrow; *Oil and Gas Journal,* November 8, 1982, 119, and February 21, 1983, 58–59; *Ghanaian Times,* November 25 and November 28, 1978; *Daily Graphic,* November 25, 1978.

Index